SECONDARY EMISSION AND STRUCTURAL PROPERTIES OF SOLIDS

VTORICHNO-EMISSIONNYE I STRUKTURNYE SVOISTVA TVERDYKH TEL

ВТОРИЧНО-ЭМИССИОННЫЕ И СТРУКТУРНЫЕ СВОЙСТВА ТВЕРДЫХ ТЕЛ

SECONDARY EMISSION AND STRUCTURAL PROPERTIES OF SOLIDS

Edited by U. A. Arifov

Director, Institute of Electronics
Academy of Sciences of the Uzbek SSR
Tashkent, USSR

Translated from Russian by
Geoffrey D. Archard

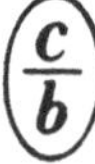
CONSULTANTS BUREAU · NEW YORK-LONDON · 1971

The original Russian text, published by FAN Press in Tashkent in 1970, has been corrected by the editor for this edition. The translation is published under an agreement with Mezhdunarodnaya Kniga, the Soviet book export agency.

У. А. АРИФОВ

ВТОРИЧНО-ЭМИССИОННЫЕ И СТРУКТУРНЫЕ СВОЙСТВА ТВЕРДЫХ ТЕЛ

Library of Congress Catalog Card Number 78-157931

ISBN-13: 978-1-4684-7214-1 e-ISBN-13: 978-1-4684-7212-7
DOI: 10.1007/978-1-4684-7212-7

A Division of Plenum Publishing Corporation
227 West 17th Street, New York, N.Y. 10011

United Kingdom edition published by Consultants Bureau, London
A Division of Plenum Publishing Company, Ltd.
Davis House (4th Floor), 8 Scrubs Lane, Harlesden, NW10 6SE, London, England

FOREWORD TO THE AMERICAN EDITION

The present collection is devoted to work carried out in the Academy of Sciences of the Uzbek SSR, in the city of Tashkent. It contains the results of recent, continuing investigations of those topics presented in my earlier books: Interaction of Atomic Particles with the Surface of a Metal* and Interaction of Atomic Particles with a Solid Surface,† the translations of which were published in English in the United States in 1963 and 1969, respectively.

In addition to theoretical and experimental work on the interaction of atomic particles with a solid surface, articles devoted to a study of atomic collisions involving multiply-charged ions, to the use of positrons for studying matter, and to structural research are incorporated.

I am happy to note the interest of foreign scientists in the work of our research team, and I hope that the present publication, together with the two works translated earlier, will promote the further development of scientific links and better mutual understanding between Soviet and American scientists.

U. Arifov, Editor

*AEC-tr-6089 (1963).

† Consultants Bureau, New York (1969).

CONTENTS

SECONDARY ION EMISSION FROM A TUNGSTEN–MOLYBDENUM ALLOY UNDER THE INFLUENCE OF SLOW IONS

D. D. Gruich, Kh. M. Khamidova, G. E. Ermakov, and U. A. Arifov

An apparatus described in an earlier paper [1] was used in order to measure the secondary ion-emission coefficients of a 50% W–Mo alloy as a function of the energy E_0 of a beam of incident Na^+ ions and the target temperature. The ion beam fell on the target at an angle of 54°. The secondary ions were separated into elastically-scattered and slow ions by the retarding-field method.

Figure 1 shows the secondary ion-emission coefficient as a function of the energy of the incident ions $K(E_0)$ for various collector potentials. The measurements were carried out on a cooling target after heating to 1400°K.

When a potential of 10 V relative to the target was applied to the collector, the latter collected all the secondary ions. By varying E_0 from 15 to 500 eV curve 1 was accordingly obtained for the $K(E_0)$ relationship (total secondary-emission coefficient as a function of primary-ion energy). Curves 2, 3, and 4 were obtained for collector potentials of +1, +2, and +6.5 V respectively; as the positive potential on the collector increased, it collected fewer and fewer of the slow ions. For a collector potential of 6.5 V the group of slow ions was completely stopped, and thus curve 4 represented the $K_{Sc}(E_0)$ relationship, i.e., the dependence of the scattered ion coefficient on the energy of the primary ions.

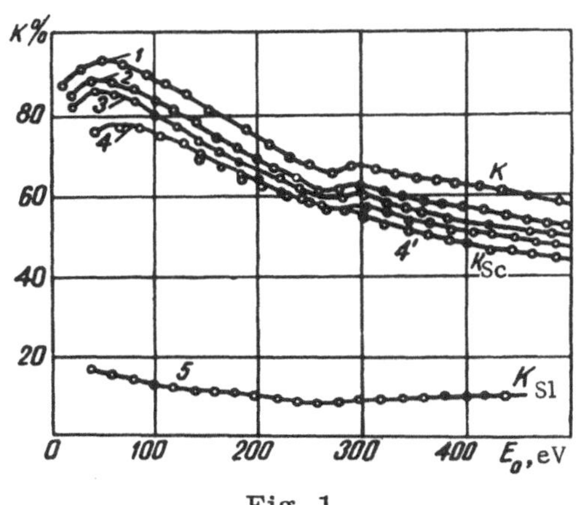

Fig. 1

The value of the collector potential completely stopping the group of slow ions was determined by reference to the width of the slow-ion peak on the energy spectra obtained in the same apparatus by means of a 127° cylindrical condenser. Figure 2 presents the oscillograms of such energy spectra obtained by bombarding the W–Mo alloys with Na^+ ions of various energies. Apart from the slow-ion peak, this also exhibits two peaks of elastically-scattered ions from the alloy components, i.e., Mo and W atoms. On reducing E_0 to 200 eV, the elastically-scattered ion peaks associated with the W atoms

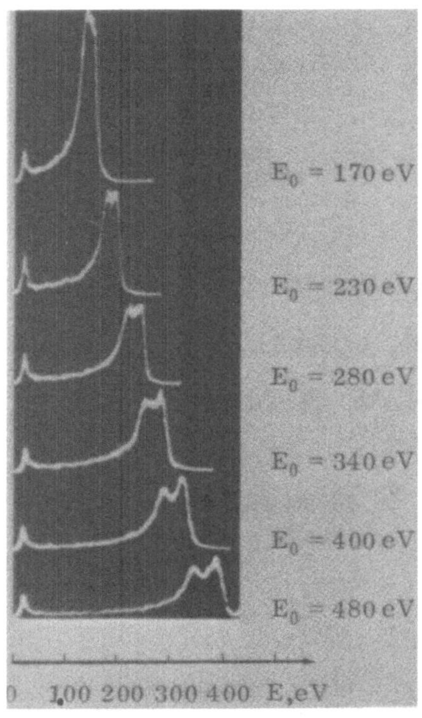

Fig. 2

vanishes; as indicated in an earlier paper [2], this is attributable to the formation of a Mo film on the surface of the W–Mo alloy on heating to a high temperature.

The $K_{S1}(E_0)$ relationship may be obtained by subtracting the $K_{Sc}(E_0)$ from the $K(E_0)$ curve.

A change in the concentration or in the thickness of a film of one of the components on the surface of the alloy should lead to a change in the ion-scattering cross section. Curves 1 to 4 (Fig. 1) exhibit a characteristic break at primary-ion energies between 300 and 350 eV; on reducing the energy of the primary ions from 500 to 350 eV, the ion scattering takes place from smaller and smaller depths in the target, the ions arising from both W and Mo atoms; for $E_0 \approx 300$ eV the depth of penetration of the ions into the target is so small that the ions are only scattered from the Mo atoms, which form a film of specific thickness on the surface of the alloy.

It was found earlier [1] that the $K_{Sc}(E_0)$ curve obtained on bombarding a tungsten target with Na^+ ions was situated at a lower level than that obtained on using the same ions to bombard a molybdenum target (see curve 4′ in Fig. 1), i.e., the scattering cross section of the Na^+ ions is greater for W than for Mo at the same value of E_0.

Thus on reducing E_0 from 500 to 350 eV, over which range the ions are scattered from both the Mo and the W atoms, there is a smooth rise in the $K(E_0)$ and $K_{Sc}(E_0)$ relationships. For E_0 values of 300 to 350 eV the ion-scattering cross section diminishes as a result of the reduction in the number of ions scattered from the W atoms, and the $K(E_0)$ and $K_{Sc}(E_0)$ curves tend to drop. For energies of under 300 eV the ions are scattered from the Mo atoms only, and the course of the $K_0(E_0)$ curve thereafter appears like that of pure Mo.

The $K_{Sl}(E_0)$ curve (Fig. 1, curve 5) shows no characteristic break in the range of E_0 between 300 and 350 eV. As indicated earlier [1], the number of slow ions depends on the ionization potential of the incident ions and the work function of the target, which remained constant in the present case; it does not depend on the energy of the incident ions.

Let us consider what would happen if, as the target temperature varied, the thickness of the Mo film on the surface of the W–Mo alloy changed. This would in fact lead to a shift in the characteristic break on the $K_{Sc}(E_0)$ curves. No such shift appeared on varying the target temperature from 900 to 1700°K. This clearly may be explained by the fact that the film thickness varied very little with temperature, and also by the fact that the $K_{Sc}(E_0)$ curve was determined experimentally for all the scattered ions emitted by the target. Such a shift in the $K_{Sc}(E_0)$ curve might well be expected for ions scattered at a particular angle Θ. In this case the $K_{Sc}(E_0)$ relationship would appear more sharply as the ions penetrated more deeply into the target.

References

1. U. A. Arifov, D. D. Gruich, and L. Yu. Chastukhina, Izv. Akad. Nauk SSSR, Ser. Fiz., No. 9, p. 1402 (1964).
2. U. A. Arifov, D. D. Gruich, and Kh. M. Khamidova, Trans. of the Eight International Conference on Phenomena in Ionized Gases, Vienna (1967), p. 20.

USE OF A THERMISTOR FOR RECORDING THE FLOW OF ATOMS WHEN STUDYING SECONDARY EMISSION

Kh. Dzhurakulov, R. R. Rakhimov, and N. V. Printseva

When using flows of neutral atoms or molecules in various physical investigations, there are two main methodological problems which have been solved: the production of a flow (flux) of particles of the desired energy, and the measurement of the corresponding intensity.

There are several methods for obtaining a flow of neutral particles [1]. The best known of these is the production of a flow of atoms by the resonance charge exchange of ions in their own gas or vapor. In this case reasonably monoenergetic flows of considerable density may be obtained over a wide energy range.

A more difficult problem is that of measuring the intensity of a flow of neutral particles. Existing methods [1-6, 12] are by no means universal; they merely supplement one another and often give a very poor accuracy. One of the simplest and most universal methods of measuring flows of atoms is the calorimetric process, which uses the thermal effect of the beam in a suitable receiver [7-8]. The accuracy of measurement in this method depends on the type of detector employed. In the majority of cases thermocouples have been used; these are only effective when working with high-density fluxes. With the development of the technique of infrared (thermal) radiation, semiconducting microthermoresistances (thermistors) have been developed; the use of these greatly extends the lower limits of the measurement of thermal radiation [7].

In this paper we develop a method of determining the flow of neutral atoms on the basis of a thermistor, and present some quantitative measurements of the electron emission of polycrystalline molybdenum bombarded with beams of argon atoms and ions in the energy range 0.5 to 5 keV. The flows of neutral atoms were obtained by the resonance charge exchange of ions in their own gas, using an apparatus described earlier [9].

The construction of the receiving part was altered in view of the introduction of the atomic-beam detector into the target region. The receiving part of the apparatus consisted of a target M, a cylindrical collector K, a protective electrode B, and an atomic-beam detector (Fig. 1a). The operating principle of the detector is based on measuring the heat developed by bombarding the surface with flows of fast particles. The principal elements are a thermoresistance (thermistor) T_1 and an atomic-beam receiver D_1. The thermistor T_1 is the measuring element; another thermistor T_2 serves to compensate the effect of the thermal vibrations of the surrounding medium. We used miniature (diameter approximately 0.1 mm) thermistors of the open type (MT-57) with an ohmic resistance of 9000 Ω at 20°C, a working temperature range from −70 to +150°C, and a temperature coefficient of 3.78%.

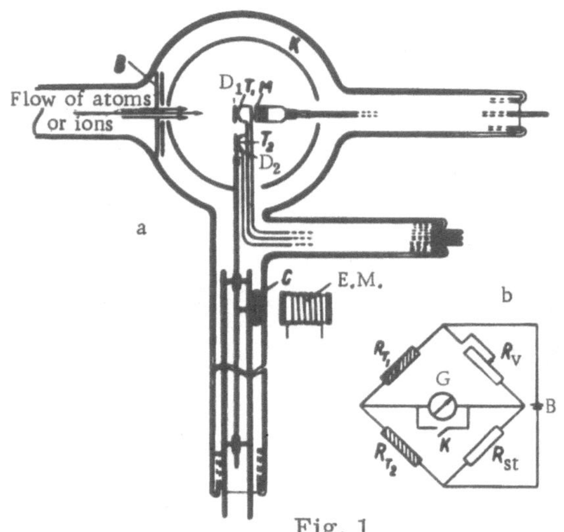

Fig. 1

The receiver for the beam of atoms was a platinum disc D_1 8 mm in diameter and 6 μ thick; the choice of platinum as receiver was governed by its comparatively low specific heat. The thermistors had good thermal contact with the receiver D and were attached to the side opposite to the beam with epoxy resin.

The detector was mounted on a separate stand and was capable of being moved by means of a special device actuated by an external electromagnet. When calibrating and recording the flow of atoms during measurements of secondary emission, the detector was placed directly behind the target at a distance of 4 to 5 mm, thus measuring the flow of atoms actually falling on the target.

When degassing the target, the detector was removed to a distance of the order of 10 cm from the collector in order to prevent severe overheating and consequent damage.

As measuring circuit, operating in conjunction with the thermistor, we used an ordinary unbalanced bridge, with the thermistors T_1 and T_2 in two of the arms (Fig. 1b). The flow of atoms was determined by reference to the degree of disbalance of the bridge as measured by the galvanometer G (sensitivity $2.9 \cdot 10^{-10}$ A/mm). The detector was calibrated by reference to a flow of ions of known energy and density, assuming that the thermal effect was the same for both ions and atoms, other conditions being equal. The assumption was based, firstly, on the fact that the inert-gas ions were almost completely neutralized at a certain distance from the target surface and then met the surface of the metal in the form of atoms [10], and, secondly, on the fact that the exchange of kinetic energy was the same on bombarding the surface with ions and atoms of the same type [11].

We obtained families of calibration curves (not illustrated in the present paper) for various energies and intensities of the bombarding ions; these showed a linear dependence on beam density.

The experiments showed that the detector could record a minimum ion flow with an intensity of 10^{-9} A at an energy of 500 eV, i.e., a flow with a power of $5 \cdot 10^{-7}$ W. The target was a molybdenum strip ($25 \times 10 \times 0.02$ mm) carefully degassed by the direct passage of a current. Degassing of the target was achieved by gradually raising its temperature to 2100°K; then it was heated for 35 to 40 h with continuous evacuation of the system at a vacuum of 1 to $2 \cdot 10^{-7}$ mm Hg. Immediately before the measurements the target was degassed by "flashing" at 2300°K. The secondary-electron emission was measured on a target heated to 1300°K.

Figure 2 shows the relation between the secondary-electron emission coefficient of a clean surface of polycrystalline molybdenum as a function of the energy E_0, on bombarding with argon atoms and ions. For energies of under 1 keV, the value of the coefficient γ for the neutral atoms was insignificant, while for ions it was approximately 10%. The electron emission for ions in this energy range is to be regarded as being due to the potential energy of the incident ions. On increasing E_0 above 1 keV, the coefficients of ion–electron (γ_i) and atom–electron (γ_k) emission rose linearly. There was a certain difference in the slopes of the $\gamma(E_0)$ characteristics relative to the E_0 axis for ions and atoms. The ratio $\frac{d\gamma_i}{dE_0}/\frac{d\gamma_k}{dE_0}$ lay around 1.1, whereas it was earlier found to be equal to 1.5 for the Ar^+–Mo and Ar^0–Mo systems [13].

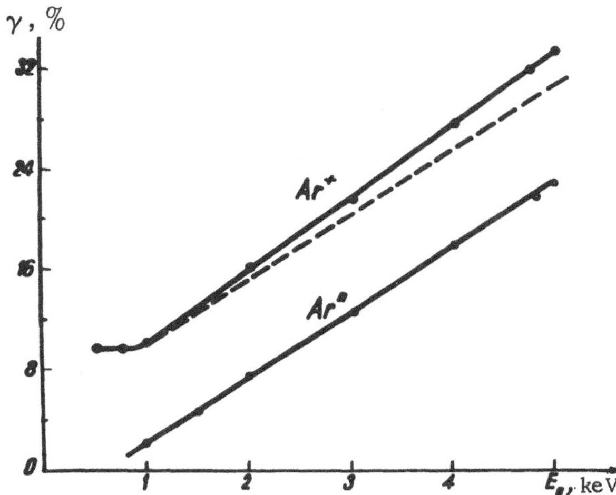

Fig. 2. Secondary-electron emission coefficient as a function of the energy of the bombarding argon ions and atoms.

In measuring the secondary emission under the impact of ions, we ourselves allowed for the number of atoms formed in the gas leaking from the ion source, which accompanied the flow of ions to the target. Furthermore, the arrangement of the detector enabled us to measure the flow of atoms actually falling on the target, i.e., we excluded possible scattering of the flow of atoms after passing the charge-exchange chamber. Allowance for these factors reduced the slope of the $\gamma_i(E_0)$ curve relative to the E_0 axis and increased that of the $\gamma_k(E_0)$ curve. Another important reason for the different inclinations of the $\gamma(E_0)$ curves to the E_0 axis found in the earlier experiments [13] lies in the difference between the state of the surfaces employed for the experiments with ions and atoms respectively. The data presented in the other paper [13] were in fact obtained for ions and atoms at different times and under different experimental conditions.

Thus the potential electron-emission coefficient depends little on the kinetic energy of the ions in the energy range studied, which supports our results obtained by the current-difference method [9, 12].

References

1. V. B. Leonas, Usp. Fiz. Nauk, 82:287 (1964).
2. D. MacClown and J. Cuevas, Pribory Nauchn. Issled., 34:95 (1963).
3. A. I. Akishin, V. S. Vasil'ev, A. F. Tulinov, and L. I. Tseplyaev, Izv. Akad. Nauk SSSR, Ser. Fiz., 28:138 (1964).
4. J. Brink, Pribory Nauchn. Issled., 37:41 (1966).
5. R. S. Woods and I. V. Fann, Pribory Nauchn. Issled., 37:102 (1966).
6. S. K. Allison, J. Cuevas, and M. Garcia, Rev. Sci. Inst., 31:1193 (1960).
7. V. F. Derbo, Thermo- and Photoresistance [in Russian] (1961).
8. V. V. Afrosimov, R. N. Il'in, V. A. Oparin, et al., Zh. Éksp. Teor Fiz., 41:1048 (1961).
9. U. A. Arifov, R. R. Rakhimov, and Kh. Dzhurakulov, Radiotekhn. i Élektronika, 8:209 (1963).
10. Sh. Sh. Shekhter, Zh. Éksp. Teor. Fiz., 7:750 (1937).
11. N. N. Flyants, U. A. Arifov, and A. Kh. Ayukhanov, Radiotekhn. i Élektronika, 8:311 (1963).
12. U. A. Arifov, R. R. Rakhimov, and Kh. Dzhurakulov, Dokl. Akad. Nauk Usbek SSR, 9:7 (1961).
13. D. B. Medved, P. Mahadevan, and J. K. Layton, Phys. Rev., 129:2086 (1963).

POTENTIAL EMISSION OF ELECTRONS FROM THE (110) AND (111) FACES OF A MOLYBDENUM SINGLE CRYSTAL UNDER ION BOMBARDMENT

U. A. Arifov, R. R. Rakhimov, and Kh. Dzhurakulov

It is already well known that the work function of the target is of particular importance in the potential emission of electrons [1, 2]. However, the dependence of the potential electron-emission coefficient on the work function of the metal has as yet been insufficiently studied; results have only been obtained for polycrystalline materials of different physical natures [3-5] and films of indefinite structure and work function [6, 7].

In order to study the manner in which the potential emission of electrons depends on the work function of the target (φ), the most suitable test objects are the different faces of a single crystal; these are greatly used at the present time for the study of thermionic emission and other surface phenomena [8].

In this paper we shall study the potential emission of electrons (potential-emission coefficient γ_π and energy distribution of the electrons) for the (110) and (111) faces of a molybdenum single crystal on bombarding with He^+ and Ar^+ ions in the energy range 0.2 to 1.2 keV. The choice of the (110) and (111) faces of a molybdenum single crystal as targets was inspired by the great contrast in their work functions.

The experiments were carried out on the apparatus described earlier [9]. The main components of the apparatus were an ion gun and a receiving section consisting of the test target and a spherical collector (diameter 90 mm). The ion gun yielded inert-gas ion beams with energies of 200 eV or over, these being monochromatic (in terms of energy) to within ± 10 eV. The conditions under which the ions were formed in the source excluded the possibility of metastable or doubly-charged ions passing into the beam.

The samples were cut from large single-crystal bars obtained by zone recrystallization [10]. The face specified was identified by x-ray structural analysis to an accuracy of 1°; the resultant single-crystal face was ground and then polished electrochemically in an electrolyte of H_2O_4 and CH_3OH, taken in a ratio of 1:7 [11]. The polishing was continued until the surface appeared perfectly ideal at the maximum magnification of the MIM-8 microscope.

The single-crystal targets had the form of a disc (diameter 11 mm, thickness 0.8 mm), and were fixed to the bottom of a cylinder made of tantalum sheet. Inside the cylinder was a tungsten spiral for indirectly heating the carrying cylinder and target.

The transverse diameter of the target was related to the diameter of the spherical collector as 11:90 ÷ 1:8. This gave an accuracy twice as great as that given by the system of electrodes used earlier [3, 5] when measuring secondary-electron energy-distribution curves.

The principal parameter was the volt–ampere characteristic (VAC) of the ion–electron emission. The VAC was measured automatically by means of circuit incorporating an electronic automatic recorder of the ÉPPV-60M type.

A saw-tooth voltage was applied between the target and collector; this was taken from a variable wire resistance, the axle of the resistance being connected to that of the recording strip in the automatic recorder. The velocity of the strip was chosen in such a way as to ensure that the useful part of the ion–electron emission VAC should be recorded within 6 sec of the disconnection of the high-temperature flash. The energy distribution of the electrons was found by numerical differentiation of the VAC [12].

The Earth's magnetic field and the stray magnetic field from the ion source were compensated in the neighborhood of the target by means of a pair of Helmholtz coils. The work function of the target was determined by measuring the thermionic emission (Richardson straight-line method). The contact potential difference between the target and collector was obtained from the retardation curves of the thermionic electrons and checked by reference to the differences between the work functions of the two single-crystal targets.

The target was degassed in two stages: 1) The target was heated for a long time by gradually raising the temperature to 2000°K and keeping the level of vacuum within 10^{-7} mm Hg; 2) the target was heated continuously for 30 to 35 h at 2000°K, and then at 2300°K in pulses 40 sec long, with a total duration of 12 h, in a vacuum of $5 \cdot 10^{-8}$ mm Hg. The target temperature was measured with a tungsten-rhenium thermocouple of the VR5-VR20 type, sealed to the edge of the single-crystal surface.

After degassing, the values of the work function equalled $\varphi_{110} = 4.9 \pm 0.05$ eV and $\varphi_{111} = 4.14 \pm 0.05$ eV. Whereas the value of φ_{111} agreed closely with recent data [13], that of the (110) face was 0.2 eV too low. The (110) face evidently contained "spots" of other faces, owing to the difficulty of producing large single crystals with reasonably perfect structures.

The individual VAC were measured while cooling the target after brief heating (2300°K); owing to the fairly large mass of the target its temperature remained between 800 and 1400°K during the measurements.

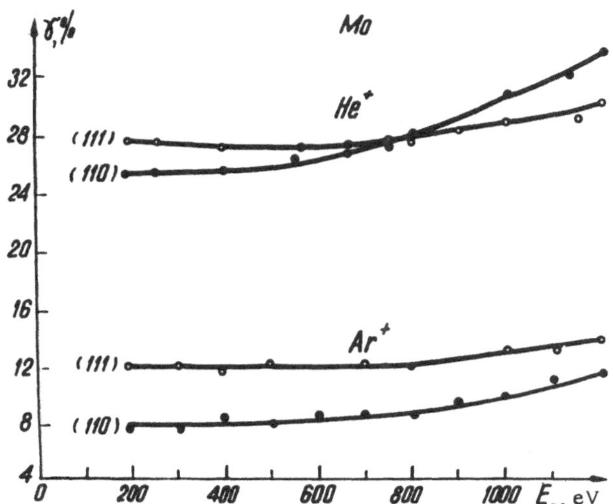

Fig. 1. Coefficient γ as a function of the energies of the argon and helium ions.

Figure 1 shows the dependence of the coefficient γ on the kinetic energy (E_0) on bombarding the (110) and (111) faces of a molybdenum single crystal with Ar^+ and He^+ ions in the energy range $E_0 = 200$ to 1200 eV. In the energy range studied the coefficients differed for the different faces of the molybdenum crystal and also for the two types of ions; the shape of the $\gamma(E_0)$ relationship was also different for different faces. Two sections may be distinguished on the curves presented.

In the first section, corresponding to a region of fairly low energies, the value of γ is independent of the energy of the bombarding particles, and the $\gamma_{110}(E_0)$ and $\gamma_{111}(E_0)$ curves run parallel to one another. The fact

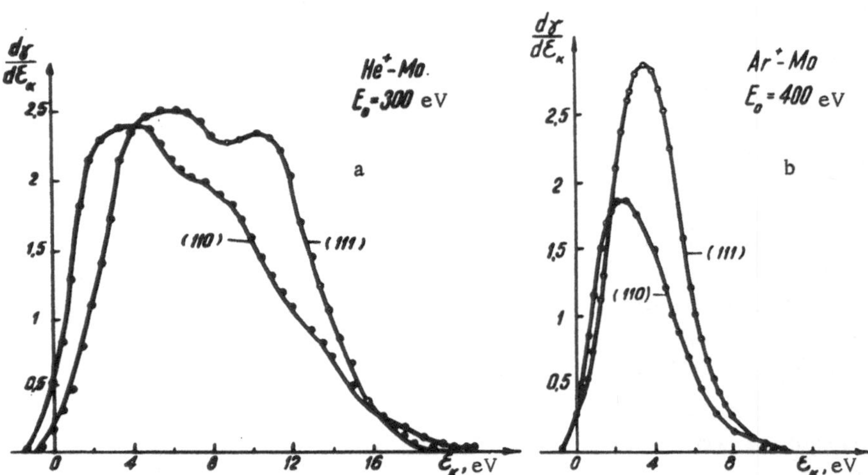

Fig. 2. Energy distribution of secondary electrons ejected from the (110) and (111) faces of a molybdenum single crystal: a) 400-eV argon ions; b) 300-eV helium ions.

that γ is independent of E_0 agrees with earlier data relating to polycrystalline targets [3, 4] and indicates that the electron emission in question is of the potential type. Thus the coefficients representing the potential emission of electrons are $\gamma_{\pi(110)} = 25.5\%$ and $\gamma_{\pi(111)} = 28\%$ for helium ions and $\gamma_{\pi(110)} = 8\%$; $\gamma_{\pi(111)} = 12\%$ for argon ions.

Whereas in the case of argon ions the value of γ_π changes by a factor of 1.5 on passing from the (110) to the (111) face, for helium ions γ_π only increases by about a factor of 1.1. This is evidently due to the fact that the electrons excited by the helium ions have a fairly hard energy spectrum; hence the probability that these will emerge into the vacuum is greater and the dependence of γ_π on φ weaker [2].

In the second section of the curves there is a monotonic rise for both types of ions and both targets. This rise is due to the superposition of an emission component associated with the kinetic energy of the ions on the potential emission of the electrons.

The energy-distribution curves of the electrons associated with potential emission on bombarding with 300-eV He^+ ions are shown in Fig. 2a for the (110) and (111) faces. The horizontal axis gives the kinetic energy of the electrons ε_k and the vertical axis gives $d\gamma'/d\varepsilon_k$, where γ' is the ratio of the number of secondary electrons with energies between ε_k and $\varepsilon_k + d\varepsilon_k$ to the number of ions falling on the target (in %). The results obtained for 400-eV Ar^+ ions are presented in Fig. 2b.

The results show that the crystallographic direction in question has a considerable effect on the energy distribution of the electrons emitted by potential emission. The change in the energy spectrum on passing from face (111) to face (110) shows no simple dependence on the work function: On passing from face (111) to face (110) there is a displacement of the maximum of the electron energy spectrum in the direction of smaller ε_k amounting to about 2.2 eV for He^+ and 1.3 eV for Ar^+. The maximum energy of the electrons emitted remains almost unaltered. In addition to this, there is a slight change in the form of the electron energy distribution for He^+.

The changes in the value of γ_π and in the energy spectrum of the electrons with crystallographic direction for a particular ion are evidently due to the difference in the degree of reflection of the electrons excited within the metal from the two types of potential barrier at the metal–vacuum interface characteristic of the two faces.

In the electron energy-distribution curves given for helium, two maxima may clearly be distinguished (particularly for the (111) face); the shape of the curve representing the electron energy spectrum is similar to that of semiconducting targets [16]. This shape of energy spectrum was explained earlier [17] as being due to the different density of states of the electrons in the valence band of a semiconductor. However, our own curves for He^+ (Fig. 2a) were obtained at a higher energy (E_0 = 300 eV) than the earlier ones [16], and the second maximum from the left of the electron energy-distribution curve is evidently due to a change in the form of the energy spectrum with increasing ion energy. Special experiments with helium-ion velocities close to the thermal value will have to be carried out in order to resolve the nature of these maxima.

References

1. Sh. Sh. Shekher, Zh. Éksp. Teor. Fiz., 7:750 (1937).
2. H. D. Hagstrum, Phys. Rev., 96:336 (1954).
3. H. D. Hagstrum, Phys. Rev., 96:325 (1954).
4. U. A. Arifov and R. R. Rakhimov, Izv. Akad. Nauk SSSR, Ser. Fiz., 24:657 (1950).
5. F. M. Propst and E. Lüscher, Phys. Rev., 132:1037 (1963).
6. U. A. Arifov, D. A. Tashkhanova, and R. R. Rakhimov, Dokl. Akad. Nauk Uzbek. SSR, 9:13 (1963).
7. U. A. Arifov, D. A. Tashkhanova, and R. R. Rakhimov, Radiotekhn.i Élektronika, No. 2, 294]1963).
8. G. N. Shuppe, Electron Emission of Metallic Crystals [in Russian], Izv. SAGU, Tashkent (1959).
9. U. A. Arifov, R. R. Rakhimov, and Kh. Dzhurakulov, Radiotekh. i Élektronika, 8:299 (1963).
10. W. Pfann, Zone Melting [Russian translation], Metallurgizdat, Moscow (1963).
11. L. Popilov and L. Zaitseva, Electropolishing and Electrolytic Etching of Metallic Microsections, Metallurgizdat, Moscow (1963).
12. M. Blanter, Methods of Studying Metals and Analyzing Experimental Data [Russian translation], Metallurgizdat, Moscow (1952).
13. G. N. Shuppe, Izv. Akad. Nauk SSSR, Ser. Fiz., 30:1935 (1966).
14. C. E. Carlston, G. Magnuson, and P. Mahadevan, Phys. Rev., 139:A729 (1965).
15. E. S. Mukhamadiev and R. R. Rakhimov, Izv. Akad. Nauk SSSR, Ser. Fiz., 30:892 (1966).
16. H. D. Hagstrum, Phys. Rev., 119:940 (1960).
17. H. D. Hagstrum, Phys. Rev., 122:83 (1961).

ELECTRON EMISSION OF THE (110) AND (111) FACES OF A MOLYBDENUM SINGLE CRYSTAL ON BOMBARDMENT WITH HELIUM AND ARGON IONS AND ATOMS

U. A. Arifov, R. R. Rakhimov, and Kh. Dzhurakulov

Earlier [1] when studying the electron emission from the (110) and (111) faces of a molybdenum single crystal bombarded with helium and argon ions in the energy range $E_0 = 0.2$ to 1.2 keV we noted that the coefficient of potential emission of the electrons γ_π exhibited anisotropic properties.

The electron emission obtained from different faces of metallic crystals bombarded with ions of higher energy (up to 10 keV) was studied elsewhere [2, 3]. It is well known that in this range of energies both potential and kinetic electron emission occur [4, 5]. Hence in order to discuss the nature of the anisotropy we must know the values of the coefficients γ_π and γ_k (γ_k is the kinetic emission coefficient) in the specified range of E_0. This can only be achieved by making a comparative study of the ejection of electrons from individual faces of single crystals by both ions and atoms of the same type of inert gas. In addition to this, investigations of this kind will reveal the dependence of γ_π on the velocity of the ion for targets with more perfect crystal structures.

In this paper we shall present the results of a study of electron emission (quantitatively expressed as the secondary-emission coefficient and the energy distribution of the secondary electrons) carried out by bombarding the (110) and (111) faces of a Mo single crystal with helium and argon ions and atoms in the energy range $E_0 = 0.2$ to 0.5 keV. The experiments were carried out in the experimental apparatus described elsewhere [5]. The vacuum conditions and the target-degassing procedure were described in our earlier paper [1]. The intensity of the neutral atoms was measured by the calorimetric method also mentioned earlier [6].

The state of the surface of the crystals was monitored by reference to the thermionic work function, which before measurement of the ion–electron emission equalled $\varphi_{110} = 4.90 + 0.05$ and $\varphi_{111} = 4.14 + 0.05$ eV.

Figure 1a shows the relationship between the electron-emission coefficient for the (110) and (111) faces of a molybdenum single crystal and the energy of the bombarding argon ions and atoms. Analogous curves for helium ions and atoms are presented in Fig. 1b. The experimental points constitute the averages of several measurements. The greatest deviation from the mean was about 4% in the case of ions and 6% in the case of neutral atoms.

The results show that the crystallographic direction influences the ejection of the electrons. In the case of the (110) face the $\gamma(E_0)$ curve slopes more steeply than the $\gamma(E_0)$ curve for the (111) face in the case of both ions and atoms. The $\gamma(E_0)$ curves obtained for the ions

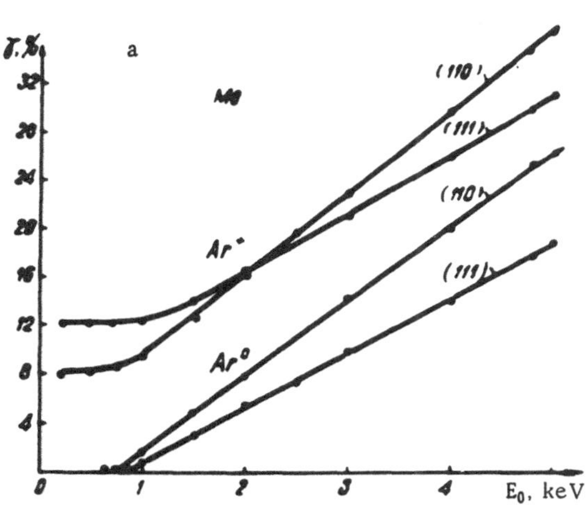

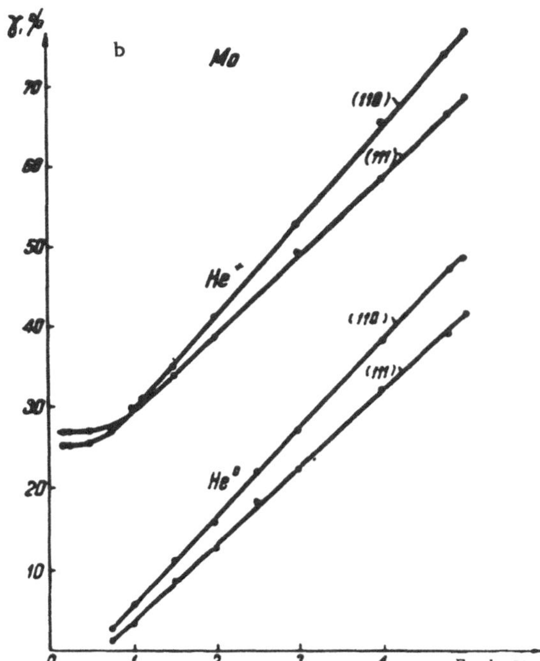

Fig. 1. Secondary electron-emission coefficient for the (111) and (110) faces of a molybdenum single crystal in relation to the energy of the ions and atoms: a) for argon; b) for helium.

intersect at $E_0 = 0.8$ keV in the case of helium. This intersection is due to the emission of electrons by virtue of the kinetic energy of the incident particles, which is higher for the face with the greater packing density of the atoms.

On bombardment of the target with atoms (instead of ions), so that only kinetic emission of electrons occurs, the anisotropy in the coefficient of kinetic electron emission γ_k for different crystallographic directions is more sharply evident. We see from the curves that the $\gamma_k(E_0)$ relationship is steeper for the (110) direction than for the (111) in the case of both argon and helium. Furthermore, the threshold energy of the bombarding ions at which the kinetic emission of electrons begins is lower for the (110) face than for the (111).

In order to elucidate the effect of the velocity of the ions on the potential-emission coefficient, it is interesting to compare the slopes of the curves representing the coefficients of ion–electron emission $\gamma_i(E_0)$ and atom–electron emission $\gamma_k(E_0)$ relative to the E_0 axis for one particular kind of gas. Taking the ratio of the angle between the $\gamma(E_0)$ curve and the E_0 axis for Ar^+ ions to the corresponding angle for Ar^0 atoms from Fig. 1a, we obtain

$$\frac{d\gamma_{i\,(110)}}{dE_0} \Big/ \frac{d\gamma_{k\,(110)}}{dE_0} \approx 1.1 \quad \text{and} \quad \frac{d\gamma_{i\,(111)}}{dE_0} \Big/ \frac{d\gamma_{k\,(111)}}{dE_0} \approx 1.06$$

and in the case of helium (from Fig. 1b):

$$\frac{d\gamma_{i\,(110)}}{dE_0} \Big/ \frac{d\gamma_{k\,(110)}}{dE_0} \approx 1.05 \quad \text{and} \quad \frac{d\gamma_{i\,(111)}}{dE_0} \Big/ \frac{d\gamma_{k\,(111)}}{dE_0} \approx 1.03$$

i.e., the ratios differ little from unity. These data are in good agreement with the results obtained for polycrystalline molybdenum, and indicate that the coefficient of potential electron emission depends very little on the velocity of the ion in the energy range studied.

Figure 2a, b presents the energy-distribution curves of electrons emitted from the (110) and (111) faces of a molybdenum single crystal bombarded with helium and argon atoms at energies of $E_0 = 1.5$ and 5.0 keV. The results obtained for bombardment with 0.3- and 5.0-keV helium ions are shown in Fig. 3a and those for 0.4- and 5.0-keV argon ions in Fig. 3b.

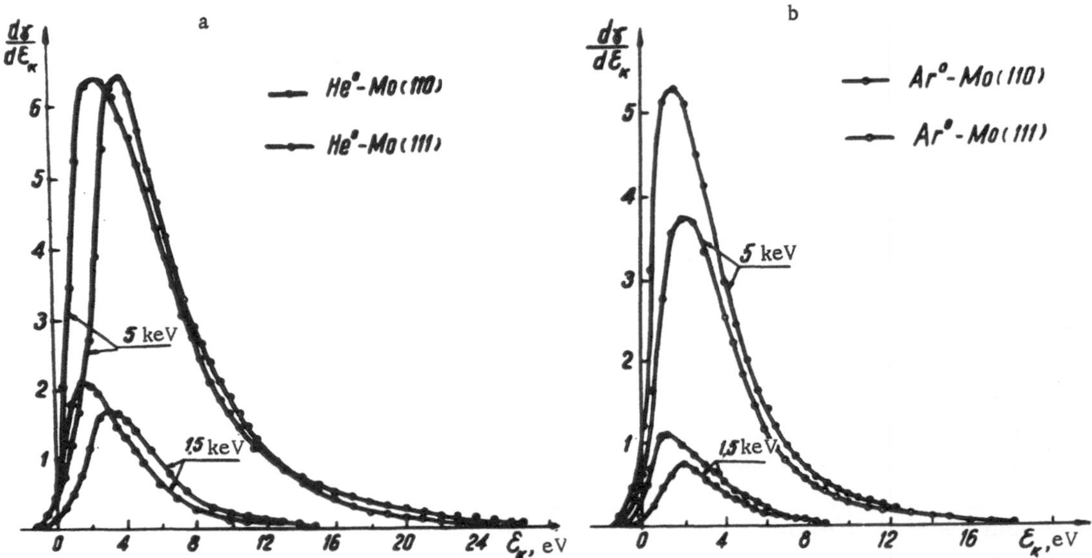

Fig. 2. Energy-distribution curves of the electrons emitted from the (100) and (111) faces of a molybdenum single crystal under bombardment by 1.5- and 5.0-keV neutral atoms: a) for helium; b) for argon.

The curves obtained for the atoms represent the energy distribution of electrons emitted by virtue of the purely kinetic effect, while, as already mentioned, the ion curves incorporate both kinetic and potential emission.

We note that the character of the energy spectrum in the case of the bombardment of either face with atoms (kinetic emission) is analogous to that associated with polycrystalline molybdenum [7]. However, there is a difference in the energy spectra obtained for the same value of E_0. This lies in the fact that the position of the curve maximum moves in the small ε_k direction on passing from the (111) face to the (110), the extent of this displacement being different for the bombardment of the target by different kinds of atoms; it is 1.3 eV for helium and 0.7 eV for argon, but is almost independent of the energy E_0. In addition to this, the electron energy spectrum is considerably enriched with slow electrons, while the maximum electron energies change very little. Analogous differences in the spectra of the electrons emitted from the two faces occur for bombardment with ions.

The dependence of the energy spectrum of the electrons on the energy of the bombarding atoms under kinetic-emission conditions is very much the same for both single crystals and polycrystalline aggregates [7]. As E_0 increases, so does the number of electrons with the most probable energy, while the positions of the maxima on the curves move in the direction of larger ε_k. For example, on increasing E_0 from 1.5 to 5.0 keV, the displacement of the maxima is 1.8 eV for the He^0–Mo (111) and 0.8 eV for the Ar^0–Mo (111) system. Furthermore the relative proportion of fast electrons and the maximum energy values also increase.

In earlier papers [2, 3] the anisotropy of the kinetic ion–electron emission coefficient γ_k for different faces of single crystals was explained as being due to a difference in transparency along different crystallographic directions.

The ratio of the kinetic-emission coefficients $\gamma_{k(110)}/\gamma_{k(111)}$ obtained for the two directions in our own experiments for $E_0 = 5.0$ keV was equal to 1.40 to 1.18 for argon and helium atoms respectively. As E_0 diminished, this ratio became slightly greater, but by no more than 15%; this was apparently due to the change taking place in the collision cross section on reducing the ve-

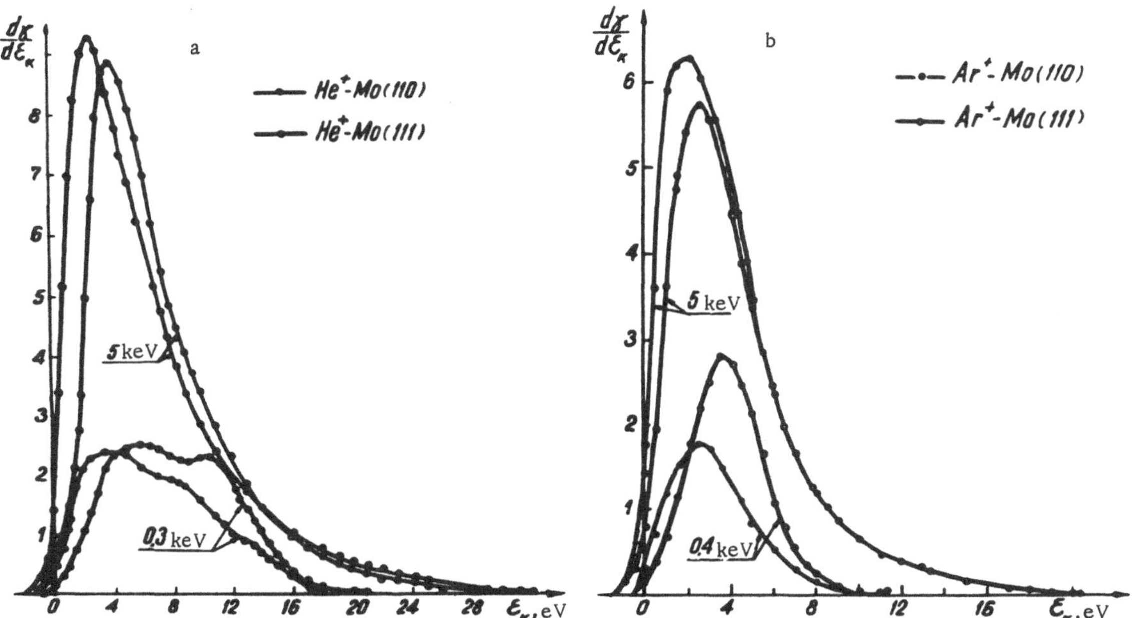

Fig. 3. Energy-distribution curves of the electrons emitted from the (100) and (111) faces of a molybdenum single crystal under bombardment by 0.3- and 5.0-keV ions: a) for helium with E_0 = 0.3 and 5.0 keV; b) for argon with E_0 = 0.4 and 5.0 keV.

locity of the bombarding particle. We see that the values of $\gamma_{k(110)}/\gamma_{k(111)}$ are different for the two types of atoms, i.e., the kinetic-emission coefficient appears more sharply anisotropic in the case of argon than in the case of helium. This is evidently due to the difference in the transparency of the crystal lattice for the two types of bombarding atoms.

The method proposed earlier [8] for determining the relative transparency of a crystal with respect to different directions for a b.c.c. lattice gives

$$\frac{\delta_{110}}{\delta_{111}} = 1.63$$

where δ_{110} and δ_{111} are the ratios of the areas covered by the atoms to the total area of the unit cell of the crystal lattice in the ⟨110⟩ and ⟨111⟩ directions respectively. To a first approximation, the number of electrons excited in the metal is proportional to the value of δ. Although there is a proportionality between $\frac{\gamma_{hkl}}{\gamma_{h'k'l'}}$ and $\frac{\delta_{hkl}}{\delta_{h'k'l'}}$, the value of the ratio $\delta_{110}/\delta_{111}$ is greater than $\gamma_{110}/\gamma_{111}$ for both argon and helium. This may apparently be explained by the difference in the conditions of excitation and egress of the electrons from the metal in the case of the two faces under consideration. The electron-emission process, in fact, comprises two stages: excitation and egress. Since in a b.c.c. lattice we have $\delta_{110} > \delta_{111}$ and $\varphi_{110} > \varphi_{111}$, while the probability that electrons (with the same energy spectrum) will leave the metal (i.e., that egress will occur) is greater, the lower the work function of the target [9], the first of these inequalities will tend to increase the ratio $\delta_{110}/\delta_{111}$ and the second to reduce the ratio $\gamma_{110}/\gamma_{111}$. We should therefore expect $(\delta_{110}/\delta_{111}) > (\gamma_{110}/\gamma_{111})$.

We found a sharper anisotropy in the γ_k coefficient than that observed earlier [2]; this was evidently due to the fact that the surface quality of the crystal used in the earlier experiments was relatively poor.

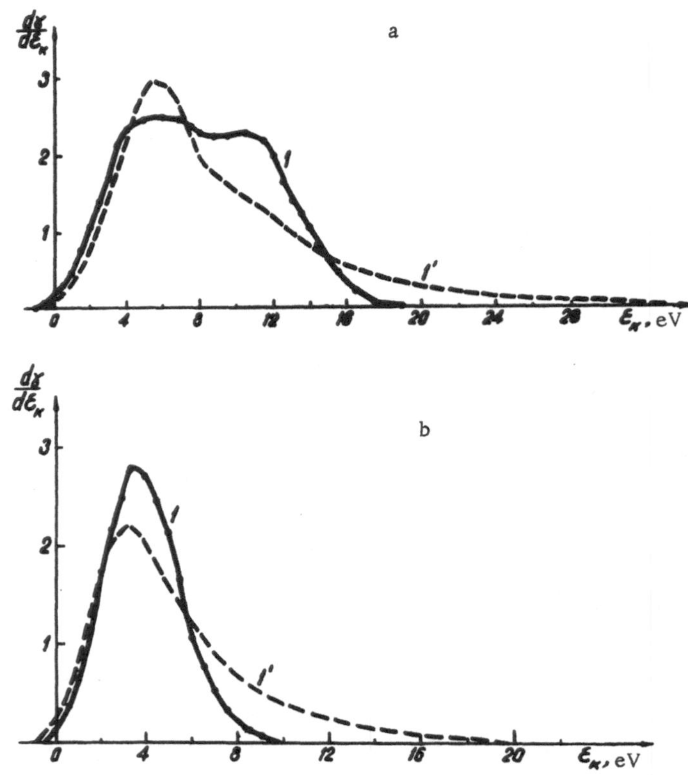

Fig. 4. Potential-emission electron energy spectrum for the (111) face: a) for helium ions with E_0 = 0.3 (1) and 5.0 (1′) keV; b) for argon ions with E_0 = 0.4 (1) and 5.0 (1′) keV.

A 1.5 times greater yield of electrons was found in another set of experiments [10] on bombarding molybdenum with ions than on bombarding with atoms. This was explained at the time as being due to an increase in the potential emission with increasing velocity of the ions. However, the ion and atom experiments were carried out at different times and under different circumstances. On remembering the strong influence of the state of the surface on ion–electron emission, we readily see that the earlier conclusions [10] are untenable.

The fact that on passing from one face to another the displacement in the position of the maximum on the electron-energy distribution curve differs from $2\Delta\varphi$ (where $\Delta\varphi$ is the difference in the work functions of the two faces), but depends on the nature of the bombarding atom (rather than the hardness of the electron energy spectrum), suggests that other factors affect the energy spectrum of the electrons as well as the work function of the target. One of these factors is evidently the ejection of slow electrons by fast Auger electrons [11], which depends on the difference in the density of the electrons in the conduction band [12] for a particular energy distribution of the electrons excited in the metal.

Let us consider the influence of the ion velocity on the electron energy spectrum for potential emission. As noted earlier, both potential and kinetic electron emission occur if the energy of the bombarding particles is high enough. Hence, in order to study the effect of the ion velocity on the electron energy spectrum for the case of potential emission, we must subtract the electron energy distribution corresponding to bombardment with neutral atoms from the electron energy distribution corresponding to bombardment with ions. Figure 4a shows the curve 1′, constituting the difference between the curves representing the He^+–Mo (111) and He^0–Mo (111) systems at E_0 = 5.0 keV. Hence curve 1′ expresses the potential-emission electron energy distribution for E_0 = 5.0 keV. For comparison, Fig. 4 also shows the energy distribution curve of the electrons emitted from the (111) face on bombardment with 0.3-keV helium ions, this curve being taken from Fig. 2a and renamed curve 1. Analogous data for argon appear in Fig. 4b.

We see from these two graphs that increasing the energy of the ions E_0 to 5.0 keV leads to the appearance of faster electrons in the spectrum, the energies of these extending up to 34 eV in the case of helium ions and 18 eV in the case of argon ions. In addition to this, there is an increase in the number of electrons lying within the spectral range ε_k = 7 to 14 eV for the helium ions, whereas in the case of argon ions the position is reversed, there being a fall in the number of electrons in the range of ε_k corresponding to the most probable energy.

Thus, in the range of energies E_0 studied, the kinetic emission of electrons is very much the same for both ions and atoms, while the potential-emission coefficient for the electrons emitted from single crystals varies only slightly with the velocity of the incident ions. However, the energy spectrum of the electrons emitted by the potential mechanism changes appreciably with ion velocity; this is due to the displacement of the energy levels of the ion close to the surface of the metal, and also to the effect of the Heisenberg uncertainty principle [9, 7].

References

1. U. A. Arifov, R. R. Rakhimov, and Kh. Dzhurakulov, Potential Emission of electrons from the (110) and (111) faces of a molybdenum single crystal under ion bombardment, this volume, p. 6.
2. C. E. Carlston, G. D. Magnuson, and P. Mahadevan, Phys. Rev., 139(3a):A729 (1965).
3. E. S. Mukhamadiev and R. R. Rakhimov, Izv. Akad. Nauk SSSR, Ser. Fiz., 30(5):892 (1966).
4. U. A. Arifov and R. R. Rakhimov, Izv. Akad. Nauk SSSR, Ser. Fiz., 24:657 (1960).
5. U. A. Arifov, R. R. Rakhimov, and Kh. Dzhurakulov, Radiotekhn. i Élektronika, 8(2):299 (1963).
6. Kh. Dzhurakulov, R. R. Rakhimov, and N. V. Printseva, Use of a thermistor for recording the flow of atoms when studying secondary electron emission (this volume, p. 3).
7. U. A. Arifov, R. R. Rakhimov, and Kh. Dzhurakulov, Izv. Akad. Nauk Uzbek SSR, Ser. Fiz.-Mat. Nauk, No. 6, p. 33 (1966).
8. G. D. Magnuson and C. E. Carlston, Phys. Rev., 129:2409 (1963).
9. H. D. Hagstrum, Phys. Rev., 96:336 (1954).
10. D. B. Medved, P. Mahadevan, and J. K. Layton, Phys. Rev., 129:2086 (1965).
11. F. M. Propst, Phys. Rev., 129:7 (1963).
12. S. Smoluchowski, Phys. Rev., 60:661 (1941).

NEGATIVE-ION SPUTTERING OF SILVER ON BOMBARDMENT WITH Cs^+ IONS

M. K. Abdullaeva and A. Kh. Ayukhanov

The mass spectrum of the negative-ion sputtering obtained by bombarding the surfaces of a number of solids with Cs^+ ions was studied earlier [1]. Very strong peaks of atomic and molecular negative ions not associated with the material of the irradiated target were observed, together with atomic and molecular negative ions of the actual target material, and also peaks associated with the ions of oxides and other chemical compounds of the target material with surface contaminants or fragments (decomposition products) of these surface compounds. The emission of the wide variety of negative ions of the first group was apparently determined by the vacuum conditions of the experimental apparatus ($P = 10^{-6}$ mm Hg) and by the intensification of the negative-ion sputtering which developed on bombarding the surfaces with Cs^+ ions.

The emission of negative atomic and molecular ions of the target material and its various compounds indicates that the surface of the target always accommodates chemical compounds; it is, in fact, well known [2, 3] that, even on a pure and degassed surface, the ion bombardment itself activates reactions between the metal and the adsorbed gases. It is also quite possible that the negative ions of the target material result from the cathodic sputtering of the principal material in the form of negative ions.

As next material for study we chose silver; this is one of a group of readily-sputtered metals [4], the atoms having a comparatively high electron affinity (2.5 eV) [5]. The targets were prepared from silver foil $50\,\mu$ thick in the form of strips 5×16 mm in area.

The sputtering chamber, with the Cs^+ ion source, was described earlier [1]. The silver was placed on the axis of the sputtering chamber. The negative ions formed on bombardment with Cs^+ ions were accelerated and directed into the analyzer chamber. The energy of the negative ions was modulated in the analyzer chamber by means of a sawtooth voltage generator. The ion currents were recorded by means of an electron-optical converter, the output of this being connected to the vertical plates of a cathode-ray oscillograph. The modulation of the ion energy, synchronized with the horizontal scan of the oscillograph, enabled a finite range of the mass spectrogram to be displayed. Different regions of these mass spectrograms were followed by changing the magnetic field of the analyzer [1].

The mass spectrum of the negative-ion sputtering achieved by irradiating a silver target with Cs^+ ions was in general similar to that obtained in the earlier work [1], although as regards composition it was richer in strong and well-resolved peaks.

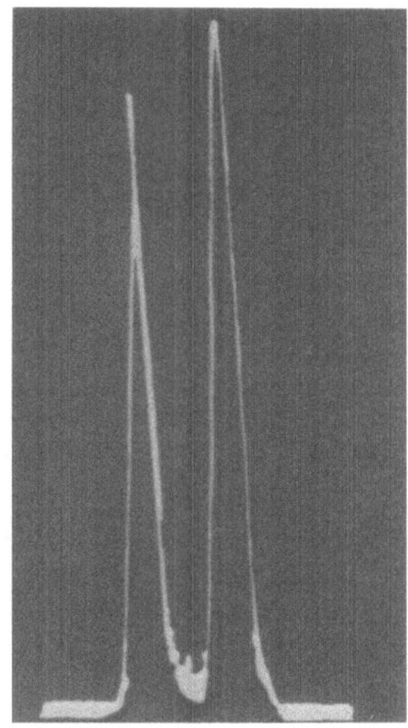

Fig. 1. Peaks of the negative Ag^- ion (107, 109).

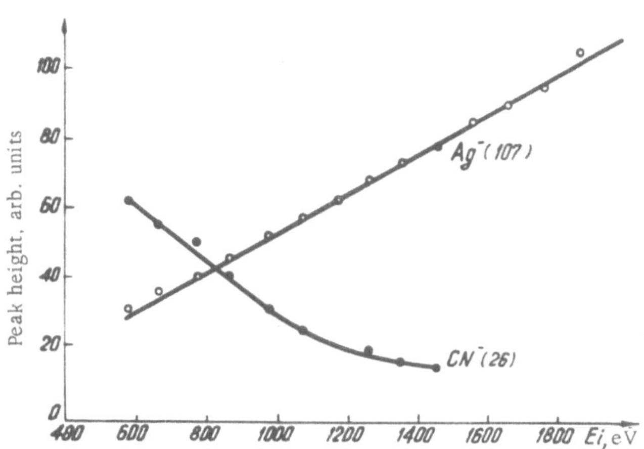

Fig. 2. Yield of negative Ag^- and CN^- ions as a function of the energy of the bombarding Cs^+ ions.

On bombarding a silver target with Cs^+ ions the negative silver ions start being ejected immediately after the onset of irradiation. The silver (107) and (109) peaks are fairly strong and excellently resolved (Fig. 1). Apart from the atomic negative ions Ag^-, we were unable to observe any other silver-containing ions in the apparatus employed.

We also determined the relation between the yield of negative Ag^- ions and the energy of the bombarding Cs^+ ions in the energy range 500 to 2000 eV for vacuum of the order of 10^{-6} mm Hg. We measured the height of the Ag^- peaks on the oscillograph screen as a function of the energy of the bombarding ions for a fixed value of the ion current. Figure 2 shows the intensity of the negative Ag^- ions in relation to the energy of the bombarding Cs^+ ions. For comparison, the same figure shows the analogous relationship for negative CN^- ions (m = 26). We see from Fig. 2 that the sputtering of silver in the form of negative ions increases linearly with increasing energy of the bombarding ions, while the sputtering of the negative CN^- ions falls to an equilibrium value with increasing energy of the primary ions in the energy range studied (0.5 to 2 keV). Other ions not containing target material behave in a similar way. The peaks of the CN^- ions were taken for comparison because of their high intensity, which greatly facilitated their observation on the oscillograph.

Similar observations were made by Bradley and Ruedl in the mass analysis of positive ions sputtered from copper [3]. The Cu^+(C) peak increased monotonically with increasing energy of the bombarding ion, this being typical of the sputtering process in general. The energy dependence of the intensity of the CuO^+ peak behaved in a different manner (Fig. 3a). Figure 3b indicates the total yield obtained by sputtering copper with argon ions [6]. There is a great similarity between curves b and c of Fig. 3, i.e., the Cu^+ yield/energy curves are similar to the sputtering characteristics.

The form of curve a (Fig. 3) was explained at the time [6] in the following way: on irradiation of a copper target with Ar^+ ions the CuO^+ peaks arise at approximately 40 eV; subsequently

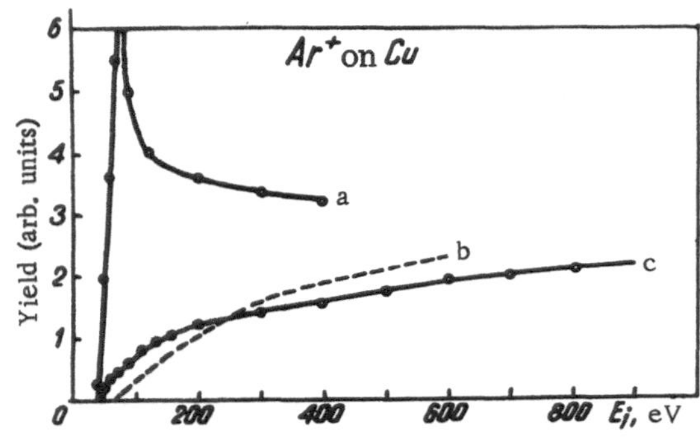

Fig. 3. Yield of Cu^+ and CuO^+ ions in relation to the the energy of the bombarding Ar^+ ions: a) CuO^+; b) Wehner's curve; c) Cu^+.

they increase sharply and resonance occurs at 75 eV. As the energy of the Ar^+ ions increases further, the peak gradually falls until an equilibrium value is attained. An electron-diffraction study of the rate of oxidation of the copper shows that Cu_2O is formed rather than CuO [7]. It was accordingly concluded that the CuO^+ ions observed constituted fragments or dissociation products of the molecules of cuprous oxide. Hence the CuO^+ and Cu^+ are not just two different types of fragment arising from the Cu_2O but particles formed by two different mechanisms: dissociation and sputtering.

It follows from a comparison of Figs. 2 and 3 that the Ag^- is not a fragment of any oxide dissociating into Ag^- and oxygen under the bombardment. Otherwise the energy dependence of the yield of Ag^- ions would have a form analogous to the energy dependence of the yield of CN^- ions or the falling part of curve a in Fig. 3.

Thus the appearance of Ag^- ions immediately after the onset of the bombardment of the silver target, the absence of molecular ions containing silver atoms, and the linear rise in the Ag^- peaks with increasing energy of the bombarding Cs^+ ions all indicate that these particles have arisen as a result of the cathodic sputtering of the principal element in the target, rather than constituting fragments of any compounds formed on the surface and sputtered by the ion bombardment.

References

1. A. Kh. Ayukhanov and M. K. Abdullaeva, Izv. Akad. Nauk SSSR, Ser. Fiz., 30(12):2000 (1966).
2. R. C. Bradley, J. Appl. Phys., 30(1):1 (1959).
3. R. C. Bradley and E. Ruedl, J. Appl. Phys., 33(3):880 (1962).
4. M. Kaminskii, Atomic and Ionic Collisions on the Surface of a Metal, Mir, Moscow (1967), Chap. 10.
5. N. S. Buchel'nikova, Negative ions, Usp. Fiz. Nauk, 65(3):354 (1958).
6. G. K. Wehner, General Mills Research Dept., Minneapolis 13, Minnesota, Ann. Rep., No. 2136 (1960).
7. A. Rohnquist and H. Fischmeister, J. Inst. Metals, London, 89:65 (1960).

NEGATIVE-ION SPUTTERING OF TUNGSTEN BOMBARDED WITH ALKALI METAL IONS

V. A. Shustrov and A. Kh. Ayukhanov

Analysis of the conditions governing earlier experiments on cathodic sputtering indicates that, even under the best vacuum conditions attainable, the desired purity of cold-irradiated surfaces is never achieved in the course of cathodic sputtering, owing to the occurrence of adsorption as well as the penetration of the primary ions. Hence some of the power of the ion beam is used up in ejecting contaminant atoms introduced both by the prinary-ion beam itself and also by the adsorption of residual gases from the main volume of the apparatus. A mass-spectrometric study of the composition of the sputtered particles [1-3] reveals molecular ions corresponding to chemical compounds formed by the target atoms with adsorbed gases. Using a radio-isotope method, we showed earlier [4] that the sputtering of tungsten in the form of negative ions might constitute a substantial proportion of the total sputtering. The mechanism underlying the sputtering of target material in the form of chemical compounds should clearly differ from that underlying its sputtering from the lattice of the pure material.

A study of negative-ion sputtering should reflect the laws governing the sputtering of the target material in the form of chemical compounds with adsorbed gases.

Our experimental method was based on the use of radioactive isotopes. A polycrystalline tungsten foil containing the radio-isotope W^{185} was subjected to sputtering.

In the earlier communication [4] we described our modifications to the radio-isotope method of studying cathodic sputtering; these firstly improved the experimental conditions and secondly enabled the particles to be discriminated in accordance with their state of charge. In order to obtain the absolute values of the cathodic-sputtering coefficient, an exact measurement of the number of positive ions falling on the target and the number of tungsten atoms ejected by these ions is required. The determination of the number of positive ions bombarding the target amounts to a measurement of the incident ion current. The incident ion current, with due allowance for secondary emission from the target, was measured by the usual method, given in detail elsewhere [5]. For measuring the amount of sputtered material we found the relation between the active and inactive atoms of the sample in an indirect manner. From the irradiated material we prepared carefully calibrated samples (standards) and counted the radioactivity pulses from these in an installation of the PC-64 type. This gave us the number of pulses per minute corresponding to 1 mg of radiometric material. Having established this relationship, it was easy to determine the amount of sputtered material received by the collector, from the number n' of pulses per minute recorded by the installation from the specified collector under the same counting conditions. The specific activity of the tungsten used in our experiments enabled us to measure an amount of material as small as 10^{-9} g.

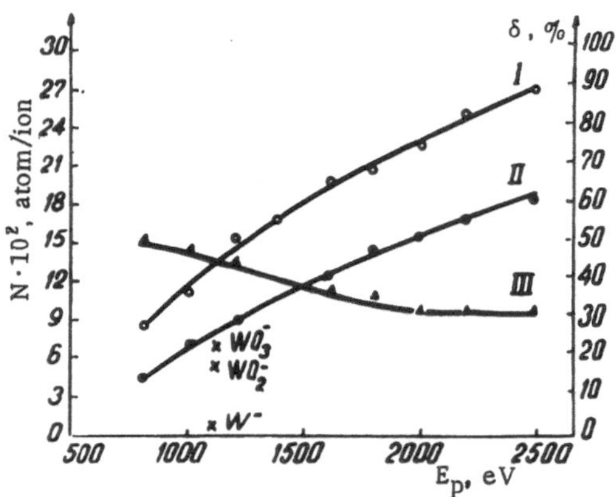

Fig. 1. Sputtering of a cold tungsten target by Cs^+ ions. I) Sputtering coefficients for $V_k = 0$, II) for $V_k = -300$ V. III) Proportion of negative ions of tungsten-containing compounds among the total amount of sputtered material (δ).

In degassing the material, the target temperature was raised to approximately 1800°K, and degassing was continued until the pressure in the apparatus reached $(5 \text{ to } 7) \cdot 10^{-6}$ mm Hg. Then the target heating was disconnected and the ion beam was made ready. The heating of the source was arranged so as to give an ion current density of 5 to 10 μA/cm². In measuring the primary-ion current, the target was short-circuited with the collector, thus eliminating errors due to secondary processes in the target–collector system. After completing the preparation of the ion beam, the target was again cleaned by heating to about 1800°K and giving five or six flashes to 2500°K. During the experiments the target remained at room temperature. The residual-gas pressure in the target region fluctuated between 1 and $4 \cdot 10^{-6}$ mm Hg under working conditions, according to the type of ions employed. Using the preparatory sequence described, the results were reproducible to within 15%.

Figure 1 shows the relationship between the cathodic-sputtering coefficient of tungsten and the energy of the Cs^+ ions. The points on curves I and II were obtained alternately in a single experiment. In the experiments with Cs^+ the collector was exposed for 5 to 10 min. Curve I was obtained with $V_k = 0$ (no potential difference between target and collector) and characterized the energy dependence of the total sputtering coefficient of a cold, degassed tungsten target in a vacuum of about $1 \cdot 10^{-6}$ mm Hg. Curve II was obtained with $V_k = -300$ V, i.e., negatively-charged sputtered particles with an energy of under 300 eV were unable to reach the collector. For a voltage of $V_k = -300$ V almost all the negative ions containing tungsten were stopped. Hence curve II describes the neutral component of cathodic sputtering. The difference between the ordinates of curves I and II corresponds to the negative-ion sputtering component. It was shown earlier [4] that this component constituted negative ions of tungsten-containing compounds. If for each specific energy we plot the ratio of the differences between the ordinates of curves I and II to the ordinate of curve I, we obtain the energy dependence of the proportion of the total sputtering associated with the sputtering of tungsten in the form of negative ions (curve III).

Krohn [3] used a mass spectrometer to establish the fact that, on bombarding tungsten with Cs^+ ions, tungsten–oxygen compounds were practically the only negative ions ejected. Krohn also gave the ratios of the heights of the main peaks in the mass spectrum. By using these data [3] we may find the relative proportions of these compounds in the total amount of sputtered material. Unfortunately the data in question only related to one ion energy (E_p = 1100 eV). Figure 1 illustrates points corresponding to the three tungsten-containing negative ions in the earlier data: WO_3^-, WO_2^-, WO^-.

We see from Fig. 1 that in the ion energy range $E_p \approx 1000$ eV approximately half the ejected particles constitute negative ions of tungsten–oxygen compounds, the oxygen being adsorbed from the main volume of the apparatus at a residual gas pressure of 10^{-6} mm Hg. In order to elucidate the role of residual gas pressure, we carried out some special experiments in which a titanium getter was employed in order to improve the vacuum. The target treated in the ordinary way was sputtered at $8 \cdot 10^{-6}$ mm Hg with a constant energy of the primary ions. After this the titanium getter was sputtered, and the pressure was thus reduced to $2 \cdot 10^{-7}$ mm Hg. Further measurements were made at the new level of vacuum. The experiments showed that a change in pressure of approximately 1.5 orders of magnitude had

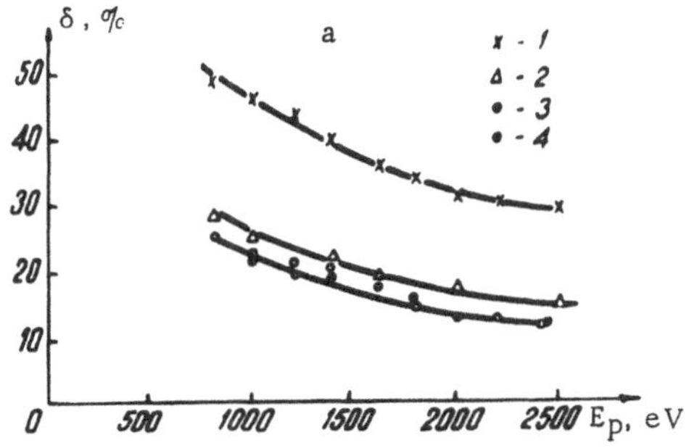

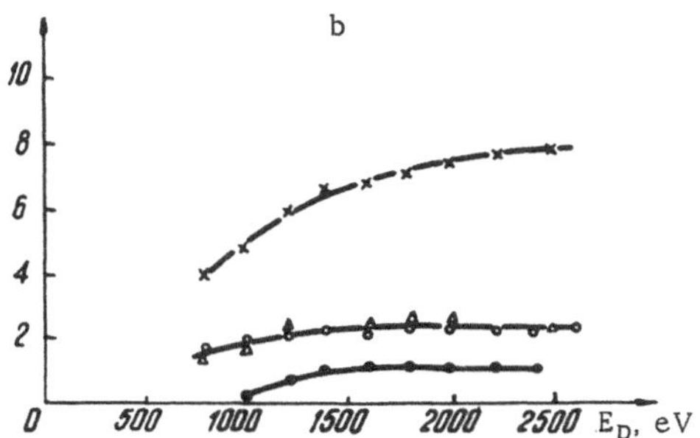

Fig. 2. Relation between the energy of the alkali ions and: a) ratio δ; b) negative-ion sputtering coefficient N^-: 1) Cs^+; 2) Rb^+; 3) K^+; 4) Na^+.

little effect on δ; the sequence of changes in the energies of the bombarding particles was also unimportant. This was to be expected, since the contamination of the surface under the vacuum conditions described took place in a time much shorter than that corresponding to the exposure of one collector (5 to 10 min).

Figure 2a shows the energy dependence of δ for four alkali ions. For the Rb^+, K^+, Na^+ ions the proportion of negative ions in the total sputtering differs comparatively little, while for Cs^+ ions it is almost twice as large and for primary-ion energies of E_p = 800 eV constitutes half the total sputtering. With increase of E_p the value of δ falls for all the ions. Starting from an energy of about 2000 eV there is a tendency for δ to become independent of E_p. The relative change in δ for the range up to E_p = 2000 eV is approximately the same for all the ions and lies between 0.3 and 0.4.

The form of the energy-dependence curve $\delta(E_p)$ may be explained qualitatively in the following manner. The number of atoms of the alkali metal x adsorbed by unit surface of tungsten is determined by the current density of the ions x^+, the accommodation coefficient of the atoms x, and the sputtering coefficient of the adsorbed atoms x by the primary-beam ions x^+. The primary-ion current density may be regarded as constant during the experiments. The accommodation coefficient of the atoms of the alkali metal x is clearly proportional to $(1 - K_p)$, where K_p is the scattering coefficient of the ions x^+. In the energy range under consideration, K_p falls slowly with increasing energy of the primary ions [5], so that the accommodation coefficient in this range of energies of the ions x^+ slowly increases. The sputtering coefficient of the film of x atoms irradiated with x^+ ions increases with rising energy of the incident ions. The combined influence of these factors means that the degree of coverage of the tungsten with the adsorbed atoms falls with increasing primary-ion energy in the range $E_p < 2000$ eV. This is accompanied by an increase in the work function of the bombarded surface and a reduction in the proportion of negative ions in the overall composition of the sputtered material. For $E_p > 2000$ eV, dynamic equilibrium sets in between the adsorption and ejection of alkali metal atoms, the work function remains almost constant, and the value of δ therefore changes very little with further increases in the energy of the bombarding ions.

The coefficient of negative-ion sputtering may easily be found from the experimental data in Fig. 1. For this purpose we have to find the ratio of the number of ejected negative tungsten-containing ions to the number of Cs^+ ions which have fallen on the target in the same time. In practice this reduces to finding the difference between the ordinates of curves I and II. Here we neglect the ejection of particles containing more than one tungsten atom.

Figure 2b shows the absolute values of the negative-ion sputtering coefficient N^- for the ions Cs^+, Rb^+, K^+, Na^+. We see that N^- coincides for Rb^+ and K^+. In the range $E_p < 1600$ eV, N^- increases with increasing E_p, changing by about 35% as E_p doubles. For $E_p > 1600$ eV the negative-ion sputtering coefficient is independent of the energy of the primary ions. The coefficient N^- changes analogously for Na^+ ions; only the absolute values of the coefficients are half those corresponding to the K^+ ions. The curve $N^- = f(E_p)$ incorporates no saturation in

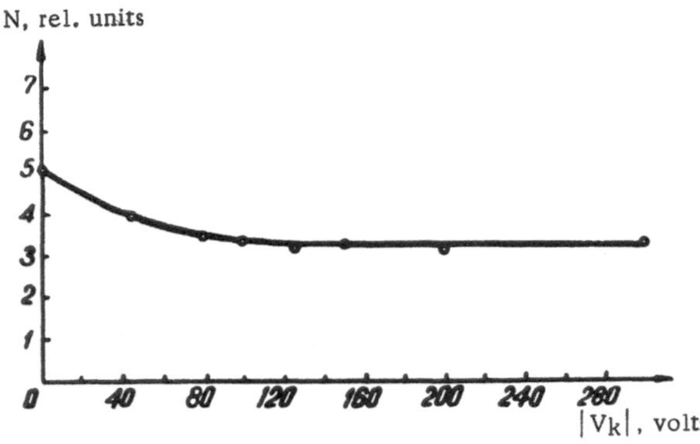

Fig. 3. Amount of sputtered material received by the collector as a function of the retarding potential V_k.

the case of Cs^+. In the range of ion energies considered the absolute value of N^- changes by more than a factor of two.

On increasing the retarding potential of the collector from zero to V_k we obtain a special kind of "retardation curve" for the negative tungsten-containing ions. The results of such an experiment with Cs^+ ions at an energy of $E_p = 2500$ eV are presented in Fig. 3. The sensitivity of our measurements (~10^{-9}g) is insufficient to enable us to discuss the limiting energies of the ejected negative ions. However, we may conclude that the number of ions with energies of $E >$ 150 eV is negligibly small compared with the total number of sputtered particles. This justifies our choice of the working retardation voltage of $V_k = 300$ V. We see moreover from the curve of Fig. 3 that the majority of the negative tungsten-containing ions have an energy of under 50 eV.

It is interesting to consider the possibility of the ejection of molecules with similar energies from the surface. In order to secure a valid solution to this problem, we must know the composition of the sputtering products, the mechanism underlying the formation of metal-containing compounds, their connection to the surface of the metal, and the mechanism underlying the transfer of momentum to the molecules of these compounds. However, the majority of these parameters are unknown and we can only discuss the broad aspects of the problem. According to the momentum theory of sputtering and the concept of pair collisions, energy passes to one of the atoms of the ejected molecule. Let us consider the WO molecule for the sake of simplicity. We may suppose [9] that the oxygen atom is only connected to one tungsten atom. Let us consider that such a molecule escapes with an energy of 100 eV, which it received when a sputtering particle struck a W atom. In order to remove the molecule as a single unit, it is essential that the energy E′ of the relative motion of the tungsten and oxygen atoms should be no greater than the bond energy of the molecule $E_b \cong 10$ eV (the bond energy of the WO_3 molecule is 9 eV [10]). In our case $E' \leq E_0 - E_c$, where E_c is the energy of the motion of the center of mass relative to the laboratory system, E_0 is the energy of the tungsten atom received in the collision. The equality sign will hold if there is no bond between the tungsten and oxygen atoms.

The energy of the motion of the center of mass may be found as follows:

$$E_c = \frac{m_W^2 v_W^2}{2(m_O + m_W)}$$

where m_W is the mass of the tungsten atom;
m_O is the mass of the oxygen atom;
v_W is the velocity of the tungsten atom relative to the laboratory system.

The energy E′ is determined thus:

$$E' \leqslant \frac{m_W v_W^2}{2} - \frac{m_W v_M^2}{2(m_O + m_W)} = \frac{m_O}{m_O + m_w} \cdot E_O \approx 8 \text{ eV}$$

The value of 8 eV is the greatest possible. Thus the coarse estimate here presented admits the possibility that a WO molecule with an energy of E ≈ 100 eV may be ejected.

A dependence of the coefficient of cathodic sputtering on the density of the ion current was noticed in a number of earlier investigations; in some [6-8] it was explained as being due to the contamination of the target surface. The absence of such a dependence, on the other hand, was regarded as a sign that the sputtered surface was clean. We carried out no experiments with gas ions; however the data which we did obtain suggested that there was a certain amount of doubt regarding the foregoing conclusions [6-8]. In order to check this we set up an experiment in which a cold tungsten target, degassed in the manner described, was bombarded with 1600-eV K^+ ions in a vacuum of $5 \cdot 10^{-6}$ mm Hg. Measurements showed that on changing the ion current density by about 100 times the sputtering coefficient remained constant. However, the sputtered surface was clearly contaminated, since in the case of $V_k = -300$ V the amount of sputtered material reaching the collector diminished considerably. It was also found that the sputtering coefficient representing the yield of tungsten in the form of negatively charged ions of tungsten-containing compounds N^- was also independent of the ion current density K^+.

References

1. R. C. Honig, J. Appl. Phys., 29:3 (1958).
2. Ya. M. Fogel', R. P. Slabospitskii, and N. M. Karnaukhov, Zh. Tekhn. Fiz. 30:7 (1960).
3. V. E. Krohn, J. Appl. Phys., 33:12 (1962).
4. V. A. Shustrov, V. I. Poltoratskii, and A. Kh. Ayukhanov, Izv. Akad. Nauk Uzbek. SSR, Ser. Fiz.-Mat. Nauk, No. 2, pp. 65-69 (1963).
5. U. A. Arifov, Interaction of Atomic Particles with the Surface of a Metal, Izd. AN Uzb. SSR, Tashkent (1961).
6. H. G. Scott, J. Appl. Phys., 33:6 (1962).
7. W. Laegreid and G. K. Wehner, J. Appl. Phys., 32(3):365 (1961).
8. M. I. Guseva, Fiz. Tverd. Tela, 1:10 (1959).
9. J. W. McBain, Sorption of Gases and Vapors by Solids [Russian translation], Gostekhizdat Fiz.-Mat. Lit, ONTI, Moscow (1934).
10. J. Kaye and T. Laby, Tables of Physical and Chemical Constants [Russian translation], Gostekhizdat Fiz.-Mat. Lit., Moscow (1962).

SECONDARY ELECTRON EMISSION FROM THE SURFACE OF GERMANIUM AND SILICON ON BOMBARDMENT WITH ALKALI IONS

É. Turmashev and A. Kh. Ayukhanov

The secondary electron emission arising as a result of the bombardment of solids with ions has been widely studied in recent years. Reliable data have been obtained as regards the true nature of these phenomena and their dependence on the many parameters of the incident ions and metal targets [1-10]. Two mechanisms governing the emission have become sharply distinguished, namely, those associated with the kinetic and potential energies of the incident ions [1]; the dependence of the corresponding coefficients on the energy and nature of the bombarding ions has been studied in detail [1, 4, 5], and a specific relationship has been established between the coefficients of ion–electron emission and the crystallographic structure of the irradiated surface. Considerable attention has been paid to the part played by the mass and velocity of the incident ions [1-3]. For this purpose the coefficients of secondary ion–electron emission have been determined over a wide range of ion masses and the so-called isotopic effect has been studied by bombarding the same target with ions of two isotopes of the same element. For the majority of the ions and targets studied, it has been established that the efficiency of the ejection of electrons increases with falling mass of the incident ion for the same ion energies. It has been shown that, over a wide range of masses, the velocity of the ion is by no means the only factor determining the efficiency of electron ejection [1, 7]. In certain cases the isotopic effect has indicated a pure velocity dependence, although deviations have still occurred [2]; these results referred only to cases in which the masses of the incident particles were smaller than those of the target atoms.

The maximum efficiency of the ejection of electrons in dielectric films (alkali halide salts) is observed when the masses of the incident ion and target atom are the same, i.e., when there is a clear relationship between the secondary ion–electron emission and the proportion of energy transferred to the target atoms in an individual act of collision [8]. The existence of such a relation is of great importance in discovering the mechanism of kinetic ion–electron emission; hence it is important to conduct further investigations into the relationship over a wide range of masses and to study the isotopic effect for both $m_1 > m_2$ and $m_1 < m_2$, not only for metals and dielectrics, but also for semiconducting materials.

For this purpose we determined the coefficient of ion–electron emission from germanium and silicon on bombarding them with alkali metal ions. The construction of the apparatus was fully described elsewhere [9]. By way of a target we used single-crystal Ge and Si plates $0.5 \times 7 \times 15$ mm in size. Before installation in the apparatus, the samples were ground, etched

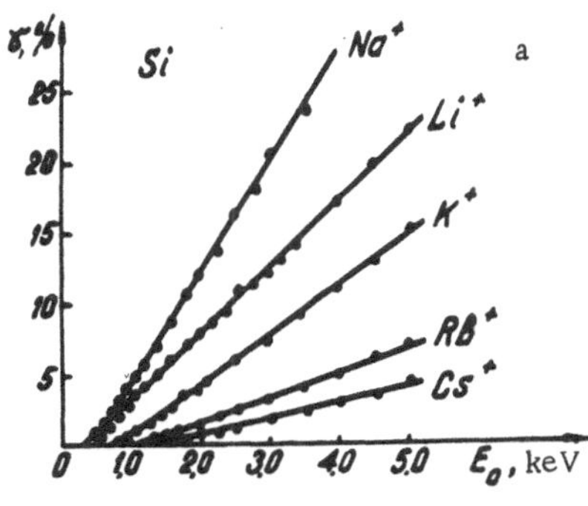

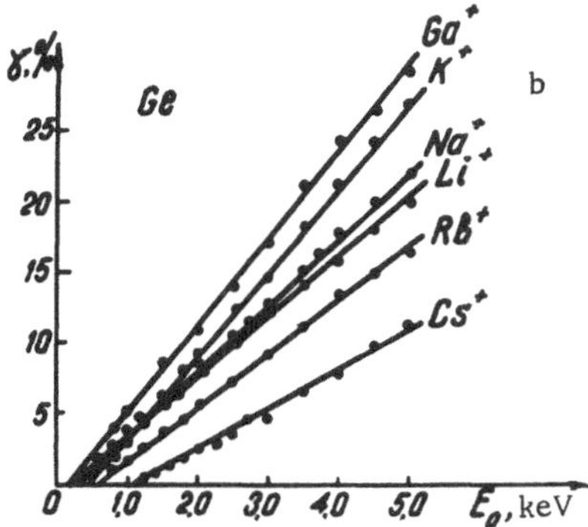

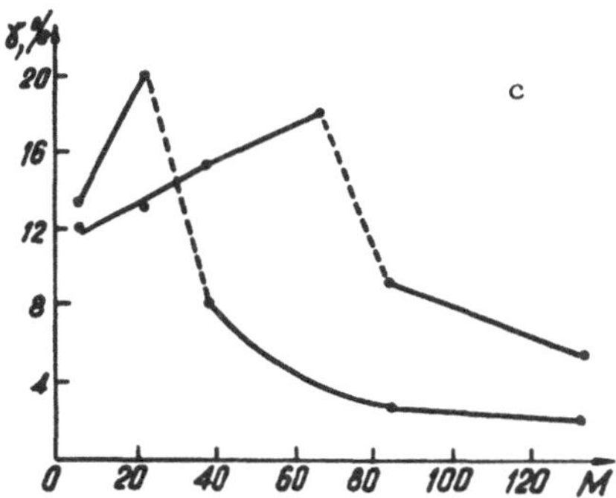

Fig. 1. Dependence of the coefficient γ on energy when bombarding Si (a) and Ge (b) with alkali ions and Ga ions, and also on the mass of the incident ions (c).

in CP-4 solution, washed with distilled water, and dried. The samples were fixed in the apparatus with tantalum clamps; the surface nevertheless exhibited a certain amount of contamination, and for making exact measurements we had to degas and clean the samples in vacuo. Both these processes were carried out by heating the samples in vacuo to as high a temperature as possible by the direct passage of an electric current. The apparatus was evacuated with mercury-vapor and getter pumps; in the processing of all parts of the apparatus the pressure was always better than $1 \cdot 10^{-7}$ mm Hg. The measurements were carried out by the double-modulation method [10], the secondary-electron currents being separated from the negative ions by means of a weak magnetic field. The intensity of the primary ion currents was 10^{-7} to 10^{-8} A/cm^2.

Figure 1a shows the energy dependence of the coefficient of ion-electron emission (γ) on energy when bombarding a clean silicon (110) surface with alkali ions. We see from the figure that, starting from a certain specific threshold value, the quantity γ increases almost linearly with increasing energy of the bombarding particles. The slopes of the curves are different for different ions. For energies above the threshold and equal energies of the ions, γ is greater when the silicon is bombarded with Na^+ ions. The energy dependence of γ on bombarding a clean germanium surface with alkali and Ga^+ ions is shown in Fig. 1b. In this case, the greatest values of γ for all the alkali ions (considering equal energies of the ions) are obtained on bombarding germanium with K^+ ions. The values of γ obtained on bombarding germanium with Ga^+ ions were greater than the corresponding values of γ for K^+.

On bombarding germanium and silicon with ions of the alkali metals, the maximum values of γ were obtained when the mass of the bombarding ions was close to that of the target atoms (Fig. 1c). However, the secondary ion–electron emission coefficient exhibited no simple dependence on the maximum energy transferred to the target atom in a single act of pair collisions. The character of the relationship between γ and the energy transferred differed for the two cases $m_1 > m_2$ and $m_1 < m_2$.

References

1. U. A. Arifov, Interaction of Atomic Particles with the Surface of a Metal, Izd. AN UzbekSSR, Tashkent (1961).
2. W. Ploch, Zs. f. Phys., 130:174 (1951).
3. K. Brunne, Zs. f. Phys., 147:161 (1957).
4. N. N. Petrov, Fiz. Tverd. Tela, 2(5):940 (1960).
5. N. N. Petrov, Fiz. Tverd. Tela, 2(6):1300 (1960).
6. I. A. Abroyan, Fiz. Tverd. Tela, 3(2):588 (1961).
7. I. A. Abroyan, M. A. Eremeev, and N. Petrov, Izv. Akad. Nauk, Uzbek SSR, Ser. Fiz., 30(5):884 (1966).

8. A. S. Smirnov and V. M. Lovtsov, Trudy Fiz.-Tekh. Inst., Akad. Nauk, Uzbek SSR 5:126 (1953).
9. U. A. Arifov, Kh. Kh. Khadzhimukhamedov, and A. I. Yunusov, Dokl. Akad. Nauk Uzbek SSR, 1:20 (1965).
10. U. A. Arifov, L. Kh. Ayukhanov, and S. V. Starodubtsev, Zh. Éksp. Teor. Fiz., 26(6):714 (1954).

SECONDARY EMISSION OF NEGATIVE PARTICLES FROM NiCr AND CuBe ALLOYS

Kh. Kh. Khadzhimukhamedov, A. Kh. Makhmudov, and A. Kh. Ayukhanov

The emission of negative particles from solids has been studied in a large number of experimental investigations. However, the nature of these particles has for a long time remained without proper consideration. It was established earlier [1] that the secondary emission of negative particles comprises electrons with a soft energy spectrum and negative ions with a hard spectrum, extending to several tens of electron volts. The secondary emission may readily be separated into two components, electrons and negative ions. For this purpose a weak magnetic field (some 50 Oe) is applied in the target region; the secondary electrons are strongly deflected by the field and fall on a special collector.

The nature and mechanism of formation of the secondary ions were also established in the earlier treatment [1]; they comprised gas ions formed as a result of cathodic sputtering. The necessary conditions for the appearance of negative ions are:

1) An insufficiently high vacuum, leading to the presence of electronegative gas atoms with a high electron affinity S in the main volume of the apparatus and on the target surface;

2) the presence of a film of electronegative atoms reducing the work function φ of the target on the object surface. Owing to the high value of S, the sputtered residual gas atoms capture electrons in the target and leave the surface as negative ions.

The emission of secondary negative and positive ions and electrons was later [2] studied by irradiating molybdenum with H^+, He^+, Ne^+, Ar^+, Kr^+ and W, Ta, Cu, Fe with He^+, Ne^+, Ar^+ ions in the ion-energy range E_0 = 10 to 40 keV. It was found that the negative- and positive-emission coefficients K_- and K_+ were of the same order, while the ion–electron emission coefficient γ was two orders of magnitude greater than K_- and K_+. The results of these experiments indicate the electronegative contamination of the target surfaces.

It was subsequently shown [3] that the value of γ and the slope of the $\gamma(E_0)$ curve diminished with increasing mass of the ion m_2. Certain authors consider that γ is independent of m_2 [4], or else that this relationship is more complicated [5] than indicated by others.

Many metallic films (Na, K, Rb, Cs, Mg, Ca, Sb, Bi) deposited on refractory substrates were irradiated with positive alkali ions in the earlier investigation [1]. For all these films except Sb and Bi, negative ions of the O^-, Cl^-, H^-, OH^-, O_2^- type and others were observed. These were apparently absent from the Sb and Bi films because of the failure to satisfy condition 2.

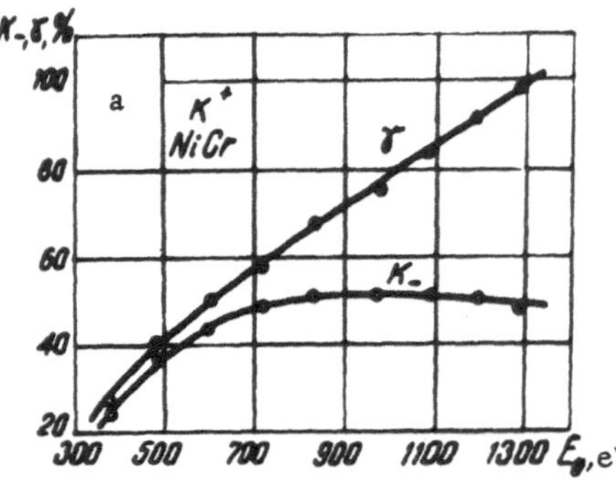

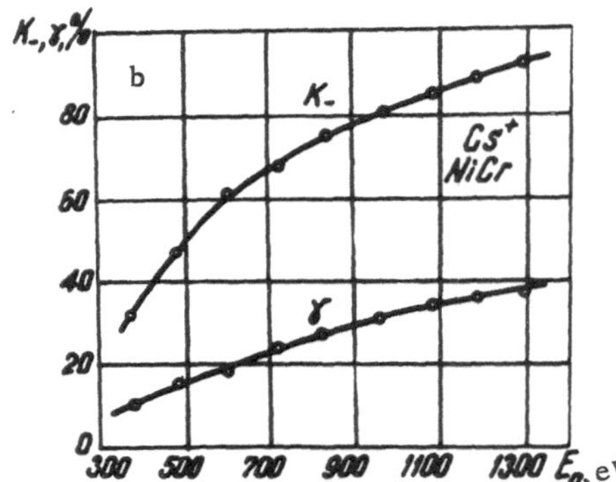

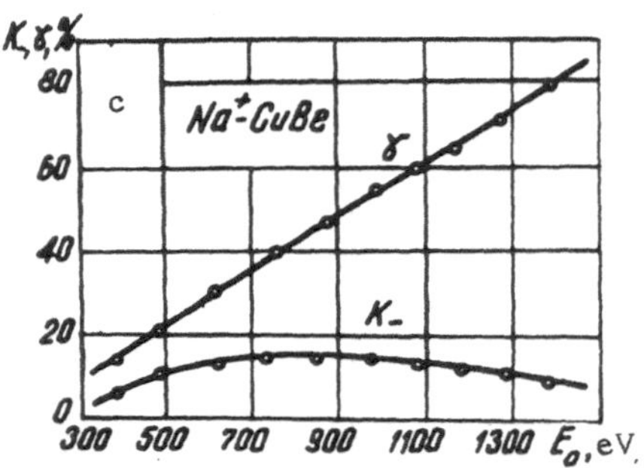

Fig. 1. Curves relating K_- and γ to the energy of the bombarding K^+ ions on NiCr (a); Cs^+ on NiCr (b); Na^+ on CuBe (c).

Secondary electron–electron emission from alloys was studied most fully in the later papers [4-5], alloys such as Cu–Be, Cu–Al–Be, AgMg, NiBe, AlMg, NiZr, NiTi, and others being examined.

Ion–electron emission from alloys has been very little studied. Higatsberger, Demorest, and Nier [6], and Kuznetsov [7] studied secondary electron emission from various alloys under the influence of gas–ion bombardment, as well as Hg^+ and Hg^{++} ions, in the energy range E_0 = 0.5 to 0.6 keV. It was found that the ion–electron emission coefficient γ varied in inverse proportion to the square root of the mass m_2 [6, 7] and increased linearly with increasing E_0. However, no division of the secondary particles into negative ions and electrons was carried out in these papers, although this is important, for example, in the development of electron multipliers, such as those used as detectors in mass spectrometers. Such a separation is also important because in certain cases the coefficient of secondary negative-ion emission reaches 100% [1].

In this paper we shall present the results of an investigation into the separate secondary emission of negative particles from NiCr and CuBe alloys bombarded with Na^+, K^+, and Cs^+ ions in the energy range E_0 = 400 to 1400 eV. The experiments were carried out by the oscillograph double-modulation method [8] in an apparatus of ordinary construction. The secondary electrons were separated from the negative ions in a magnetic field and collected in a rear collector. The method enabled us to obtain the volt–ampere characteristics of secondary electron and ion emission on the screens of two oscillographs. On the basis of these oscillograms of the volt–ampere characteristics we plotted graphs of E_0 against the ion–electron emission coefficient γ and the secondary negative-ion emission coefficient K_- (Figs. 1a, b). These graphs were obtained by bombarding the NiCr alloy with K^+ (a) and Cs^+ (b) ions. The contribution of the negative ions to the secondary emission of negative particles is considerable (Fig. 1a) and in some cases even dominant (b). Hence in such investigations it is essential to allow for the secondary emission of negative ions from the effective alloy emitters used in the multipliers. The values of γ for both ions increase linearly with rising E_0.

As in the case of metals [3], so also in the case of alloys the value of γ and the slope of the $\gamma(E_0)$ curve are greater for the light ion (K^+) than for the heavy ion (Cs^+). This is because, for the same energy, the light ion has a greater velocity and ejects more electrons from the target surface. For potassium the value of K_- increases with energy and passes to saturation at $E_0 \sim 900$ eV; in the case of cesium K_- also increases with E_0 but a little more slowly at high energies.

As already mentioned, the necessary conditions for the appearance of negative ions include a reduction in the work function, associated with the adsorption of electropositive atoms, and the presence of foreign atoms with a high electron affinity on the surface (for example, oxygen, chlorine, etc.). For certain modes of activation of such alloy emitters, both conditions are satisfied to a certain extent.

In some of our experiments, conducted at an apparatus pressure of p $\sim 1 \cdot 10^{-6}$ mm Hg, the atoms adsorbed from the primary beam of alkali ions played the part of additional centers reducing the work function φ of the surface. A reduction in φ also occurs on bombarding a CuBe target with Na^+ ions (Fig. 1c), although sodium is less electropositive than cesium and potassium. Under the influence of a heavy ion the sputtering efficiency will be greater than under that of a light ion. In the case of Cs^+ the value of K_- reaches 90% at $E_0 = 1300$ eV.

It is well known [1] that secondary-ion and kinetic electron emission depend not on the charged state of the bombarding atomic particle but rather on its mass. Hence, under similar conditions of the activation of alloy emitters, analogous phenomena will clearly occur, whether their surfaces are bombarded with inert gas ions or neutral atoms. The values of the ratio K_- /γ will depend mainly on the mass of the primary ions and on the contribution of potential emission for $V_i \geq 2\varphi$, where V_i is the ionization potential of the bombarding particles.

References

1. U. A. Arifov, Interaction of Atomic Particles with the Surface of a Metal, Izd. AN Uzb.SSR, Tashkent (1961), Chap. X.
2. Ya. M. Fogel', R. P. Slabospitskii, and A. B. Rastrepin, Zh. Tekhn. Fiz., 30:63 (1960).
3. U. A. Arifov, Interaction of Atomic Particles with the Surface of Solids, Izd. Nauka, Moscow (1968).
4. H. Brunning, Physics and Application of Secondary Electron Emission [Russian translation], Izd. Sovetskoe Radio, Moscow (1958).
5. N. A. Soboleva, A. G. Berkovskii, N. O. Chechik, and R. E. Eliseev, Photoelectronic Apparatus, Izd. Nauka, Moscow (1965).
6. M. J. Higatsberger, H. L. Demorest, and A. O. Nier, J. Appl. Phys., 25:883 (1954).
7. K. N. Kuznetsov, Trudy NII MRTP SSSR, 2:36 (1957).
8. U. A. Arifov, A. Kh. Ayukhanov, S. V. Starodubtsev, and Kh. Kh. Khadzhimukhamedov, Dokl. Akad. Nauk SSSR, 124:60 (1959).

SECONDARY ELECTRON EMISSION OF NiCr AND AlBe ALLOYS

U. A. Arifov, Kh. Kh. Khadzhimukhamedov, and A. Kh. Makhmudov

The manner in which fast electrons move in a solid depends very little on the structure of the latter and differs only slightly in metals and semiconductors. The average depth $\bar{l}$ of the penetration of an electron into a solid is proportional to its energy E_0, raised to the power ~ 1.5 [1]. The probability W_0 of the excitation of secondary electrons in matter increases as the energy of the primary electrons diminishes (as the electrons slow down in the target), i.e., at the end of the primary-electron trajectory W_0 reaches its greatest value. Inelastically-reflected electrons, moving toward the surface of the solid, also participate in the excitation of a considerable number of secondary electrons.

The secondary electrons moving in the solid themselves lose energy as a result of interaction with the crystal lattice, lattice defects, and free electrons, and also in the excitation of tertiary electrons. These electrons have a certain energy at the surface of the solid, and on overcoming the potential barrier they pass out into the vacuum. The secondary electrons have a much higher energy than thermionic electrons, and suffer less from the effects of the potential barrier than these.

In the case of dielectric targets, the method of short, sharp pulses is employed [2] in order to prevent charging of the surface with primary electrons. On bombarding thin dielectric films deposited on a metal substrate with electrons, the Molter effect takes place; this leads to very large values of the secondary-electron emission coefficient σ, measured in tens and hundreds of units. However, emitters of the Molter type have the disadvantage of being characterized by very severe inertia.

Dielectric targets are used in television transmitting tubes and in special electron-beam devices employed in storage systems with an electronic memory. Different types of multipliers (photoelectron, electron–electron, ion–electron) using secondary emission in order to amplify weak currents are based on effective emitters with high values of σ. Semiconducting-film and alloy emitters may be included among these.

More detailed investigations into secondary electron–electron emission have been carried out in connection with metallic samples. The maximum value of the coefficient $\sigma = \sigma_{max}$ varies between 0.5 and 1.8 for different metals [3]. The small value of σ_{max} is due to the great concentration of free conduction electrons in metals; a considerable proportion of the secondary-electron energy is dissipated in these, and this leads to a fall in σ_{max}.

Semiconductors and dielectrics have values of σ_{max} between 1 and 30. For the majority of elemental semiconductors, the value of σ_{max} is comparatively low. High σ_{max} (10 to 30) are found in semiconducting compounds of the intermetallic type (effective-film and alloy emitters), alkali halides, and many oxides of alkali and alkaline-earth metals.

The greater the width of the forbidden band in the semiconductors, the better are their secondary-emission properties. In alloy emitters, a semiconducting layer is formed during heat treatment and the oxidation of the active component, which has a substantial oxygen affinity.

The secondary emission of nickel [4] and copper [5] alloys has been fully studied, as well as that of AlMg [6] and AgMg [7].

In this paper we shall present some fresh data relating to secondary electron–electron emission from NiCr and AlBe alloys for primary-electron energies of E_0 = 200 to 1200 eV.

The experiments were carried out by the double-modulation method [8]. Using oscillograms of the primary current and the volt–ampere characteristics of the secondary electrons, we plotted graphs of σ against E_0 for various states of the target. We used differential pumping of the source and measuring parts of the experimental apparatus. The working vacuum in the measuring part of the apparatus was equal to ~ 5 to $10 \cdot 10^{-7}$ mm Hg. By changing the sign of the electric field, the same apparatus could be used to obtain beams of electrons and positive Na^+ ions. The density of the ion current reached $j \sim 10^{-5}$ A/cm^2; for the electron current the value of j was an order of magnitude smaller.

Figure 1 shows the $\sigma(E_0)$ curves for an unactivated NiCr target bombarded with electrons at an incident angle of $\Phi = 0°$ (1), 45° (2) and for activated NiCr with $\Phi = 45°$ (3). Activation of the emitter was carried out in a working vacuum at T = 1300°K for 60 min. We see from the figure that the value of σ increases sharply in all cases with increasing E_0, reaches a maximum at E_0 = 500 eV, and then, after falling, tends to saturation at large E_0. Increasing the angle of incidence of the primary electrons by 45° causes the value of the coefficient at the curve maximum σ_{max} to rise from 3.8 to 5; after activation of the target it is increased to 5.5.

Experiments have also been reported in which the electron emission increased on treating the targets with alkali metal vapor [9].

It is interesting to study the effect of the bombardment of alloy emitters with alkali metal ions on their emission characteristics. We carried out the following experiment. An NiCr

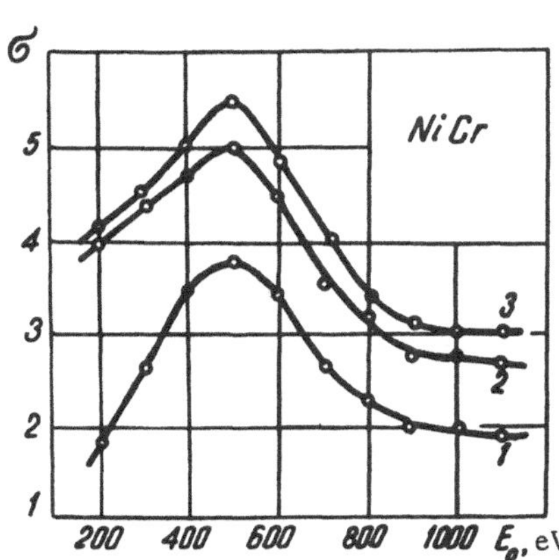

Fig. 1. Curves of $\sigma(E_0)$ for NiCr targets before (1, 2) and after (3) activation; $\Phi = 0°$ (1), 45° (2, 3).

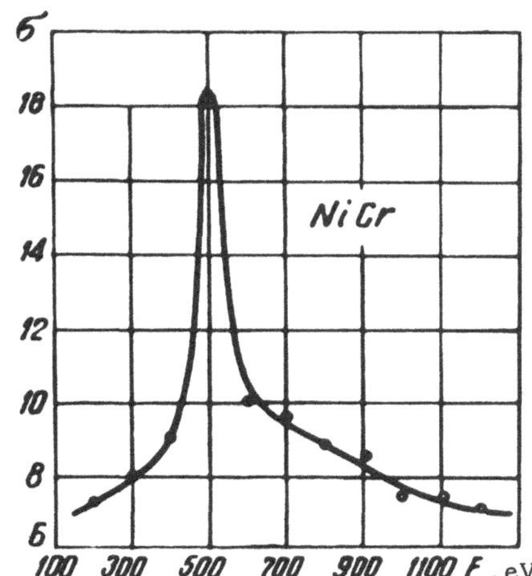

Fig. 2. Curve of $\sigma(E_0)$ for NiCr after thermal activation and bombardment with Na^+ ions.

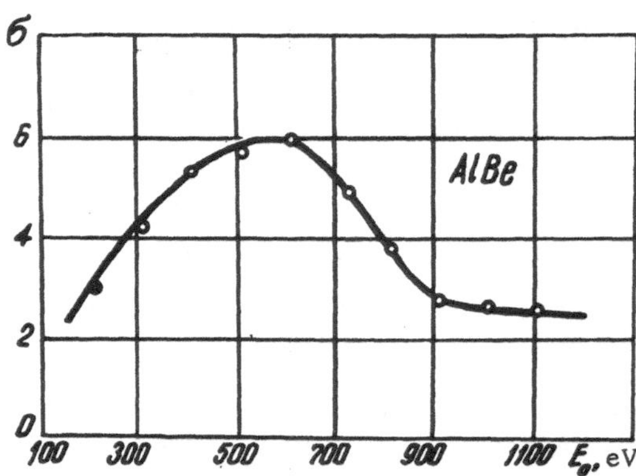

Fig. 3. Curve of $\sigma(E_0)$ for AlBe prior to activation at $\Phi = 0°$.

emitter was activated in high vacuum at 1300°K for 45 min; then it was bombarded with 500-eV Na^+ ions in the cold state for the same period. After both forms of activation an electron beam was applied to the emitter and the $\sigma(E_0)$ curve was plotted; a very high emission of the Molter type was observed, with a coefficient of $\sigma_{max} = 18$ to 20 at $E_0 = 500$ eV. One of the resultant $\sigma(E_0)$ curves is presented in Fig. 2. At the beginning and end of the curve $\sigma \simeq 7$, while in the region of $E_0 \sim 500$ eV there is a sharp maximum with $\sigma_{max} = 18.3$.

Figure 3 shows the $\sigma(E_0)$ curve for an unactivated AlBe alloy ($\Phi = 0°$). In this case the curve is very similar to those of Fig. 1. The value of $\sigma_{max} = 6$ at $E_0 \simeq 600$ eV. This alloy clearly has a better emissive capacity and its value of σ is greater in the unactivated state than that of activated NiCr (curve 3 in Fig. 1).

The presence of a maximum on the $\sigma(E_0)$ curves may be explained in the following way. While the average depth of penetration $\bar{l}$ of the primary electrons into the target is smaller than the effective depth of emission of the secondary electrons h, the value of σ will increase with increasing E_0. Since the probability of the excitation of secondary electrons increases at the end of the primary-electron range, for a depth of $\bar{l} = h$ the value of σ will reach a maximum at a certain energy value $E_0 = E_{max}$, which differs for different emitters. On further increasing E_0, the quantity $\bar{l}$ becomes greater than h, and, owing to the loss of energy by the secondary electrons in the interior of the target, the probability of their emerging into the vacuum becomes smaller. On still further increasing E_0, the value of σ begins falling. The dependence of σ on the incidence angle Φ of the primary electrons on the target may be explained in an analogous manner.

Thus, on bombarding a thermally-activated alloy emitter with positive alkali-metal ions, high values of σ may be achieved, in the same way as on treating it with alkali-metal vapor.

References

1. N. A. Soboleva, A. G. Berkovskii, N. O. Chechik, and R. E. Eliseev, Photoelectron Apparatus, Izd. Nauka, Moscow (1965), Ch. IV.
2. A. R. Shul'man and V. L. Makedonskii, Zh. Tekhn. Fiz., 22:1540 (1952).
3. H. Brunning, Physics and Application of Secondary-Electron Emission [Russian translation], Izd. Sovetskoe Radio (1958), p. 44.
4. B. S. Kul'varskaya, Izv. Akad. Nauk SSSR, Ser. Fiz., 20:1029 (1956).
5. V. N. Lepeshinskaya, Izv. Akad. Nauk SSSR, Ser. Fiz., 20:1025 (1956).
6. G. B. Stuchinskii, Fiz. Tverd. Tela, 5(3):798 (1963).
7. P. Rappaport, J. Appl. Phys., 25:288 (1954).
8. U. A. Arifov, A. Kh. Ayukhanov, and S. V. Starodubtsev, Zh. Éksp. Teor. Fiz., 26:714 (1954).
9. L. G. Leiteizen, B. M. Glukhovskoi, and E. I. Tarasova, Radiotekhnika i Élektronika, 5(12):2038 (1960).

ON THE ISOTOPIC EFFECT ASSOCIATED WITH KINETIC ION–ELECTRON EMISSION

É. S. Mukhamadiev, A. Mirzaev, and R. R. Rakhimov

Secondary-electron emission takes place in addition to other phenomena when a stream of fast charged or neutral particles strikes a solid surface. The emission of electrons from a solid may occur as a result of either the potential or the kinetic energy of the incident particles. The kinetic emission of electrons has been insufficiently studied. Up to the present time there has been no concerted opinion as to the way in which the nature of the incident particle affects electron emission, nor as to the part played by the crystal structure or work function of the target. An analysis of experimental data obtained for the secondary-emission coefficient indicates that this quantity varies in a complicated manner with the mass and state of charge of the irradiating particle. In a number of investigations, the dependence of this characteristic on the mass of the incident ion has been studied in "pure" form by using isotopic ions of one and the same substance; the influence of the structure of the electron shells in the ions on the coefficient γ is thus excluded.

Ploch and Walcher [1] bombarded Cu, Be, Mo, and Pt targets (not subjected to degassing) with singly-charged $^{6}Li-^{7}Li$, $^{20}Ne-^{22}Ne$, $^{39}K-^{41}K$ ions, and showed that, for the same velocity, both isotopes ejected the same number of electrons.

Flaks [2] also found no dependence of γ on the mass of the ion in the case of ^{6}Li and ^{7}Li ions; in the case in question the target surface was also covered with a layer of adsorbed atoms. However, Brunne [3] carried out a bombardment with isotopic Li, K, and Rb ions under better vacuum conditions and did establish the existence of an isotopic effect.

The emission of electrons from pure Mo was studied later [4] under bombardment with isotopic hydrogen ions (H_2 and D_2) in the energy range 0.1 to 45.0 keV. The isotopic effect was only observed in the case of comparatively low velocities.

The value of the coefficient γ depends on the mass ratio of the colliding particles, which determines the condition of energy transfer. Hence in the case of ion–electron emission we should certainly expect an isotopic effect, i.e., the coefficient should vary with the mass of the incident particle. However, the effective depth from which the excited electrons are ejected exerts a certain influence on the isotopic effect, and in the case of metals the effect can only be expected to be weak.

In contrast to the earlier treatments, we took account of the fairly long range of the electrons in semiconductors by comparison with metals, and in the present investigation accordingly studied the isotopic effect achieved on bombarding the (100) and (111) faces of a silicon single crystal with ^{20}Ne and ^{22}Ne ions at 0.5 to 5.0 keV.

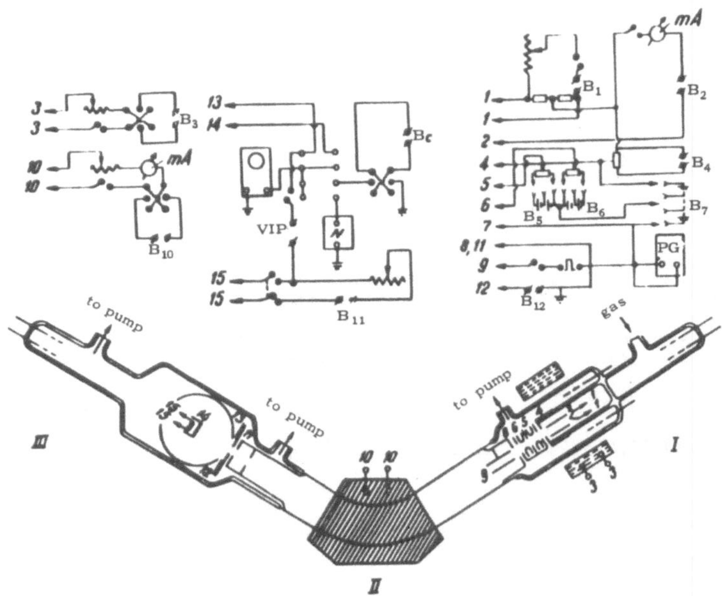

Fig. 1. Arrangement of experimental apparatus and electrical measuring circuit (schematic).

The experimental apparatus (Fig. 1) consisted of three parts: the ion source I, the mass analyzer II, and the receiver III. We used an ion source of the Finkelstein type [5], in which the ions were formed by electron impact between the heating filament 1 and anode 2 situated in the magnetic field of the solenoid 3. The shaped ion beam fell into the magnetic field of the analyzer; a 60° mass spectrometer with a resolving power of $m/\Delta m = 90$ enabled isotope ions with a particular charge to be directed on to the target separately from the fast, charge-exchange atoms. The receiving part of the apparatus comprised a target 14 and a spherical collector 13.

By way of targets we used the (100) and (111) faces of a p-type silicon single crystal $1 \times 8 \times 10$ mm in size.

The creation of clean semiconductor surfaces is quite a difficult problem. In order to clean the faces under consideration we heated them for a long period (several days) at 1500°K in vacuo. The surface of the target was considered as being clean enough when the $\gamma(E_0)$ relationship became reproducible. The secondary currents were measured by an oscillograph method, the intensity of the primary beam being modulated by applying rectangular voltage pulses to the plates 9 (Fig. 1). The low inertia of this method greatly reduced the influence of target-surface gas adsorption on the emission of secondary electrons.

The various sections of the apparatus were bounded by plane diaphragms, which enabled a pressure drop to be created in the receiving part of the apparatus. The system was evacuated with three mercury diffusion pumps, which gave a vacuum of 1 to $2 \cdot 10^{-7}$ mm Hg in the receiving part of the apparatus. After gas was admitted into the source, the vacuum in the receiving section only worsened slightly.

Before each measurement the target was subjected to a brief "flash." Several series of curves were plotted for the isotopic ions ^{20}Ne and ^{22}Ne.

Since we were working with inert gas ions, for which the condition $eV_i > 2\varphi$ (where V_i is the ionization potential of the particle and φ is the work function of the target) was satisfied,

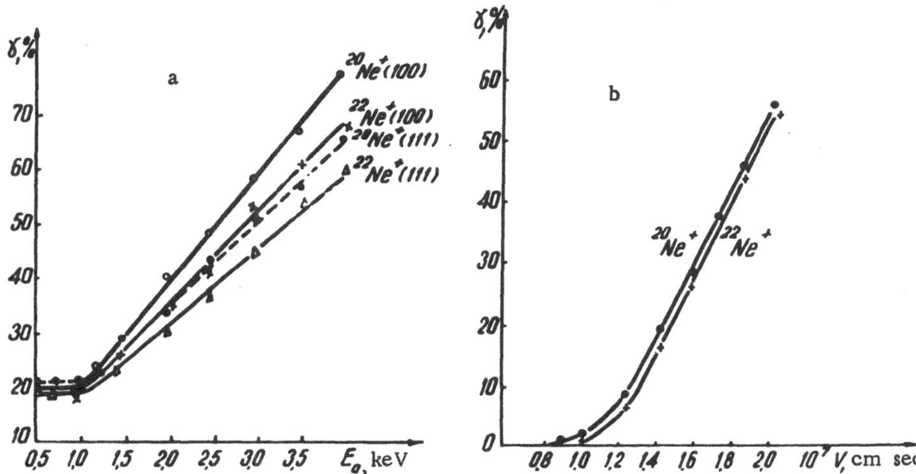

Fig. 2. Dependence of γ on E_0 for the bombardment of the (100) and (111) faces of a silicon single crystal with ^{20}Ne and ^{22}Ne ions (a), and on V for the bombardment of the (100) face of silicon with isotopic neon ions (b).

for reasonably high energies of the incident ions both potential and kinetic emission of electrons occurred. It is well known that, for neon ions up to an energy of 1 keV, potential electron emission is capable of taking place [6].

According to a number of experimental results [7, 8] the coefficient of potential emission γ_π is independent of the mass of the bombarding particle and is mainly determined by its ionization potential. Hence a determination of the value of γ_π in the energy range 0.5 to 1.0 keV should enable the potential emission to be separated from the total emission of the electrons. On the other hand, the potential emission of electrons is extremely sensitive to the state of the target surface, so that the numerical value of γ_π may be used as a measure of target surface purity. The value of γ_π in the energy range below 1 keV is very close to the value of γ obtained for silicon in far better vacuum conditions [9], being equal to about 20%.

The energy dependence of the total emission coefficient γ under bombardment by isotopic ^{20}Ne and ^{22}Ne ions is presented in Fig. 2a. As we should expect, in the low-energy range the values of γ corresponding to the two isotopes are exactly the same, both for the (111) and for the (100) face. The equal values of γ for the (100) and (111) faces in the range of potential electron emission may clearly be explained by the fact that the difference in the corresponding work functions is only about 0.06 eV [10]. The values of γ are shown in relation to ion velocity in Fig. 2b. These curves correspond to the kinetic emission only, the values of γ'_π (potential emission) having been subtracted from the total emission coefficient. We see from Fig. 2b that the coefficient γ_k is also different for the ^{20}Ne and ^{22}Ne isotopes, even when the ion velocity is the same. For example, for an ion velocity of $v = 1.23 \cdot 10^7$ cm/sec, which corresponds to an ion energy of $E_{20} = 2000$ eV, $E_{22} = 2200$ eV, the difference in γ is 2.5%

With increasing ion velocity the difference $\Delta\gamma$ becomes smaller. In order to explain the results obtained we may make use of the Davies data [11] relating to the isotopic effect when ions penetrate into a solid. Davies studied the depth of penetration of the radioactive ^{22}Na and ^{24}Na isotopes into aluminum and observed that, owing to the different degree of retardation in the solid, the light isotope penetrated to a smaller depth than the heavy one. This means that the light isotope loses more energy than the heavy in a single collision.

Hence in the surface layer of a solid the light isotope will excite more electrons than the heavy, and this will lead to different values of γ for the isotopic ions. On increasing the velocity of the ions, some of the acts of ionization penetrate into the lower regions of the solid, from which the excited electrons are unable to pass out into the vacuum. The relative contents of the isotopes in the surface layer are now almost the same; hence for high ion velocities the values of γ are the same for both isotopes. We see from Fig. 2b that the values of the coefficient for the (100) and (111) faces differ in respect of the slopes of the curves; the γ for the (100) face has a greater slope than that for the (111) face. This may be explained by the different packing density of the crystal along the [100] and [111] directions. If we consider the lower-lying levels of silicon atoms (diamond-type lattice), the greatest packing density is along the [100] direction. The ratio of the atom-covered part of the area to the total area is expressed in the following way for the (100) and (111) faces:

$$\delta_{100} = \frac{8\sigma}{a^2}, \qquad \delta_{111} = \frac{12\sigma}{\sqrt{3}\,a^2} = \frac{7\sigma}{a^2}$$

where σ is the effective cross section of the atom, and a is the lattice constant.

The ratio of the packing densities $\delta_{100}/\delta_{111}$ equals

$$\frac{\delta_{100}}{\delta_{111}} = \frac{8\frac{\sigma}{a^2}}{7\frac{\sigma}{a^2}} = 1.14$$

For an energy of the main isotope ions $E_0 = 3$ keV, the values of γ are $\gamma_{100} = 38.5\%$, $\gamma_{111} = 31\%$, and their ratio is

$$\gamma_{100}/\gamma_{111} = \frac{38.5}{31.0} = 1.24$$

By comparing the ratios $\delta_{110}/\delta_{111}$ and $\gamma_{100}/\gamma_{111}$ we see that the anisotropy in γ may be explained by the different transparency of the crystal along the [100] and [111] directions, i.e., by a change in the probability that the incident particles will collide with lattice atoms.

References

1. W. Ploch and Walcher, Rev. Sci. Inst., Vol. 22, No. 12 (1951).
2. I. P. Flaks, Author's abstract of Candidate's Dissertation, Leningrad Pedagogical Institute (1954).
3. C. Brunne, Z. Phys., 147:161 (1957).
4. U. A. Arifov, R. R. Rakhimov, M. K. Abdullaeva, and C. Gaipov, Izv. Akad. Nauk SSSR, Ser. Fiz., 26, 1103, 11 (1962).
5. Finkelstein, Rev. Sci. Inst., 11:94 (1940).
6. H. D. Hagstrum, Phys. Rev., 89:244 (1953).
7. H. D. Hagstrum, Phys. Rev., 96:325 (1954).
8. R. R. Rakhimov, Author's abstract of Candidate's Dissertation, Tashkent (1958).
9. H. D. Hagstrum, Phys. Rev., 119:3 (1960).
10. K. Heiland, Usp. Fiz. Nauk, Vol. 89, No. 4 (1965).
11. I. A. Davies et al., Canad. J. Chem., Vol. 39 (1961).

ANGULAR DEPENDENCE OF THE ENERGY SPECTRA OF SECONDARY IONS SCATTERED BY A SINGLE CRYSTAL

U. A. Arifov and A. A. Aliev

We showed in an earlier series of papers [1-5] that, on bombarding single crystals with ion beams, the general anisotropy of the angular distribution was accompanied by a marked structure in the energy distribution of the secondary ions, the nature of this being determined by the multiple collisions of the incident ion with the ordered distribution of atoms in the single crystal. The results of these investigations were in good agreement with the calculated data presented elsewhere [6], which related to a quantitative estimation of the probability of two-fold collisions taking place in addition to single collisions. However, the energy spectra indicate the occurrence of scattered ions with energies considerably greater than those of ions experiencing two-fold collisions. The presence of such ions, with maximum energies of E_m, is apparently attributable to collisions of higher multiplicity.

In this paper we shall consider the angular and energy dependence of the maximum energies of secondary ions scattered by a molybdenum single crystal as a result of one, two, and more collisions.

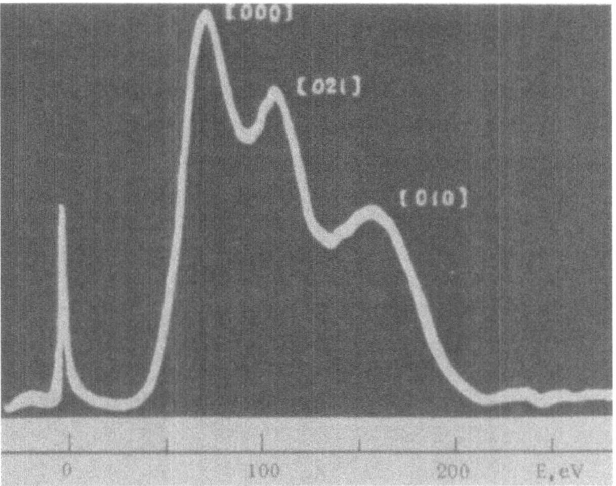

Fig. 1. Oscillogram of the secondary-ion energy distribution obtained on bombarding the (100) face of a molybdenum target heated to 1700° K with 600-eV Rb^+ ions ($\Phi = 50°$, $\Theta = 50°$).

The investigation was carried out in the apparatus described earlier [7]; the secondary-ion energy analyzer was an electrostatic condenser of the Hughes–Rozhansky type. Certain improvements and modifications in structure were incorporated in order to allow for the fact that the work was being carried out with single crystals, and also in order to increase the resolving power of the apparatus [1].

The oscillogram of the secondary-ion energy distribution (Fig. 1) was obtained by bombarding the (100) face of a molybdenum single crystal heated to 1700° K with 600-eV Rb^+ ions, the angle of incidence of the primary ions Φ being equal to 50°. Secondary ions traveling at an angle of $\Theta = 50°$ were subjected to energy analysis. The incident and scattered beams lay in a plane passing through the [010] axis of the molybdenum single crystal.

We see from the oscillogram that, as in the case of the bombardment of the (100), (110), and (111) faces of a tungsten target with alkali metal ions [1-5], so also in the present case the high-energy part of the spectrum exhibits peaks with energies greater than those of the singly-scattered ion peak. The peaks with the greatest energies correspond to ions which have been doubly scattered at atoms in the [010] and [021] directions, while the peak lying close to the peaks of the slow and evaporated ions represents the energies of ions which have experienced single collisions at the [000] atom. As origin of coordinates we arbitrarily take the [000] atom in the (100) plane corresponding to the first collision; the remaining indices on the peaks denote the atom with which the repeated collision has taken place (Fig. 2).

On reducing the energy of the primary ions E_0, further peaks appear between the [000] and [010]. The formation of these may be attributed to the ions scattered for a second time at atoms lying around the atom at which the first collision took place [6] (Fig. 2).

The highest-energy peak in the spectrum, the [010], falls off exponentially on the high-energy side; its width is clearly greater than the natural value. This is due to the presence of ions with energies greater than those acquired in two-fold collisions. With increasing E_0 the peaks corresponding to the energy of singly- and doubly-scattered ions move in the high-energy direction, and their position depends linearly on E_0 over the range $E_0 \geqslant 500$ eV. The energies of the primary ions E_0 for each of the resultant oscillograms of the distribution curve are given in Table 1. Columns 2 and 3 indicate the energies of the peaks corresponding to singly- and doubly-scattered ions for each oscillogram. The values of these energies increase

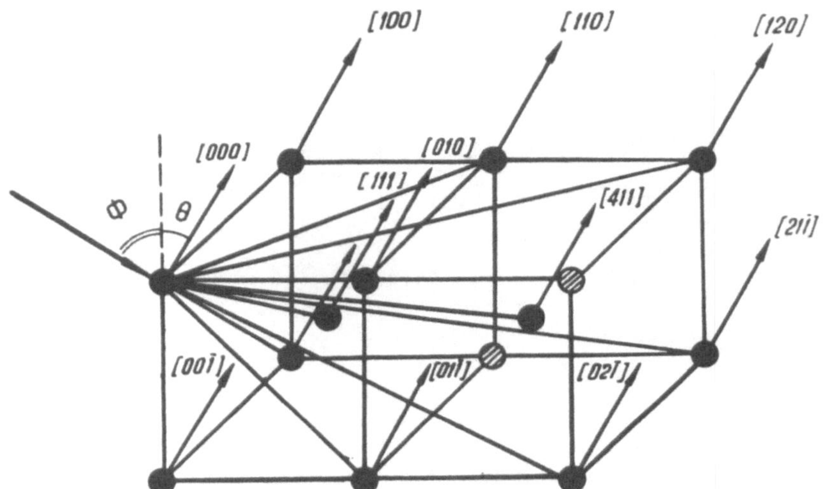

Fig. 2. Schematic representation of one- and two-fold collisions of the incident ion with the atoms of a single crystal.

with increasing energy of the incident ions. The ratios $\eta_{n_1} = \frac{E_{n_1}}{E_0}$, $\eta_{n_2} = \frac{E_{n_2}}{E_0}$, obtained experimentally are constant and equal to 0.20 and 0.40. The theoretical values of these ratios $\eta_{T_1} = \frac{E_1}{E_2}$, $\eta_{T_2} = \frac{E_2}{E_0}$ for the case of single and double collisions of the Rb^+ ions with the molybdenum target atoms are calculated from the formulas

$$E_1 = \frac{E_0(\mu - 1)^2}{(\cos\beta_1 \mp \sqrt{\mu^2 - \sin^2\beta_1})^2} \tag{1}$$

$$E_2 = \frac{E_0(\mu - 1)^4}{(\cos\beta_1 \mp \sqrt{\mu_2 - \sin^2\beta_1})^2(\cos\beta_2 \mp \sqrt{\mu^2 - \sin^2\beta_2})^2} \tag{2}$$

(where μ is the ratio of the mass of the target atom m_1 to the mass of the ion m_2, β_1 and β_2 are the scattering angles in the first and second collisions respectively) and are equal to 0.22 and 0.41; the agreement is excellent. The maximum energies of the scattered ions are indicated in column 4 of Table 1; the ratio of E_m to the incident-ion energy in the range $E_0 > 600$ eV averages 0.49.

The peaks corresponding to the slow and evaporated ions fall with increasing primary-ion energy; they assume an appreciable height for ions of low initial energies.

Figure 3a reflects the dependence of η_{n_1}, η_{n_2}, η_m (ratios of the energies of the secondary ions experiencing one, two, and more collisions to the primary-ion energy E_0) on E_0. The same figure compares these with the dependence of η_{T_1} and η_{T_2} on E_0, where η_{T_1} and η_{T_2} are the calculated ratios of K^+ ions experiencing one- and two-fold collisions with an Mo atom. The change in the dependence of η_{n_1} on E_0 is analogous to the change in η obtained for a polycrystalline sample [8-9], although it differs from the dependence of η_{n_2} and η_m on E_0 in that the values of η_{n_2} and η_m increase more rapidly than η_{n_1}, for small E_0, while the deviation in the values of η_{n_2} and η_m from the linear law starts at $E_0 \leqslant 500$ instead of 300 eV as in the case of η_{n_1}. This difference may be explained by the influence of the binding (bond) energy of the target atoms, if we consider the influence of this energy in repeated collisions as well.

Figure 4 shows the oscillograms corresponding to the energy distribution of the secondary ions obtained for different angles of incidence of the primary Rb^+ ions (energy $E_0 = 1000$ eV) on the (100) face of a molybdenum target heated to 1800°K. Oscillogram 1 (Fig. 4) is ob-

TABLE 1. Change in the Energy Spectra with Changing Primary-Ion Energy for Rb^+ on Mo(100), $\Phi = 60°$, $\Theta = 50°$

E_0	E_{n_1}	E_{n_2}	E_m	$\eta_{n_1} = \frac{E_{n_1}}{E_0}$	$\eta_{n_2} = \frac{E_{n_2}}{E_0}$	$\eta_m = \frac{E_m}{E_0}$
25	15	—	23	0.58	—	0.93
50	22	—	38	0.42	—	0.76
100	31	56	66	0.31	0.55	0.66
150	42	75	90	0.28	0.50	0.61
200	50	94	115	0.25	0.47	0.58
250	56	110	138	0.23	0.44	0.55
300	66	126	170	0.22	0.42	0.54
350	74	144	182	0.21	0.41	0.52
400	84	165	208	0.21	0.41	0.52
450	90	180	230	0.20	0.40	0.51
500	105	200	250	0.21	0.40	0.50
550	115	220	275	0.21	0.40	0.50
600	126	240	295	0.21	0.40	0.49
700	140	280	344	0.20	0.40	0.49
800	165	320	384	0.21	0.40	0.48
900	180	360	432	0.20	0.40	0.48
1000	200	400	490	0.20	0.40	0.49

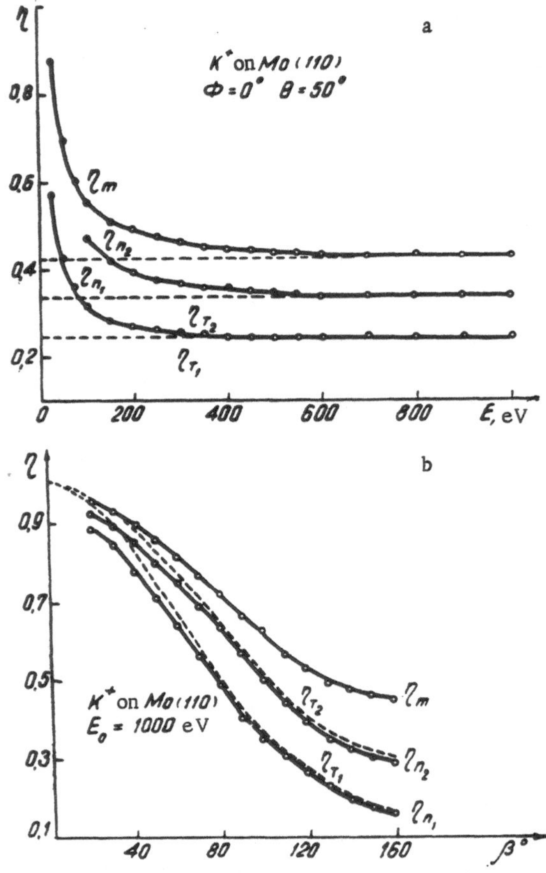

Fig. 3. Dependence of η_{n_1}, η_{n_2}, and η_m (for K^+ ions striking the 110 face of a Mo target) on the energy of the primary ions E_0 (a) and on the scattering angle β (b).

tained for an angle of incidence of the beam on the target equal to $\Phi = 30°$. Each successive picture corresponds to a 10° increase in the angle Φ. The escape angle Θ is 50° in every case. We see from the oscillograms that the peaks corresponding to single and double collisions move in the high-energy direction with increasing angle of incidence of the primary ions. Table 2 shows the ratios of the secondary-ion energies to the primary energy calculated for single and double collisions of Rb^+ ions with individual Mo atoms by means of formulas (1) and (2), and also the experimental values obtained from the oscillograms. Table 2 also shows the ratios of the maximum secondary energies to the energy of the primary ions. We notice that the values of η_{T_1}, η_{n_1} and η_{T_2}, η_{n_2} are in reasonable agreement.

We may conclude from the oscillograms and Table 2 that, with increasing angle of incidence of the primary-ion beam, the maximum energies of the secondary ions tend to approach the value corresponding to two-fold collisions. Moreover an increase in the angle Φ leads to an increase in the intensity of the peaks corresponding to the doubly-scattered ions as compared with the intensity of the single-scattering peak. In some cases ($\Phi = \Theta \geqslant 80°$) there is an absence of the single-scattering peak, which no longer appears on the background of the peaks corresponding to doubly-scattered ions. This effect, i.e., the apparent vanishing of the single-scattering peak, clearly indicates a sharp rise in the number of repeatedly colliding ions when the angle of incidence and the escape angle approach glancing values. In the present case the incident and escaping beams lie in a plane passing through one of the low-index axes of the Mo crystal. Hence in the case of glancing incidence the ions are deflected through small angles in the scattering plane before falling into the analyzer as a result of successive collisions with atoms in the surface chains parallel to this axis. This is confirmed by analyzing a model of the reflected ions arising from a chain of atoms in the surface of a single crystal as the incident angle approaches the glancing value [10-11].

The peaks of the slow and evaporated ions vary with the angle of incidence. With increasing angle of incidence of the primary ions, the heights of these peaks diminish, and for $\Phi \geqslant 80°$ they vanish altogether. This indicates the absence of primary ions from the surface, all having penetrated into the lower levels of the target. The vanishing of the peaks corresponding to the slow and evaporated ions is clearly related to the mechanism by which target surfaces are cleaned by ion bombardment.

Figure 3b shows the dependence of the values of η_{n_1}, η_{n_2}, and η_m on the scattering angle β on bombarding the (110) face of a molybdenum target heated to 1800°K with K^+ ions at an energy of $E_0 = 1000$ eV. The broken curves correspond to values of η_{T_1}, η_{T_2} calculated for single and double collisions of K^+ ions with individual Mo atoms from formulas (1) and (2). The curves of $\eta_{n_1}(\beta)$ and $\eta_{n_2}(\beta)$ in general agree with $\eta_{T_1}(\beta)$ and $\eta_{T_2}(\beta)$, although there is a ten-

TABLE 2. Energy Spectra in Relation to Increasing Angle of Incidence of the Primary Ions

Angle of incidence	0°	10°	20°	30°	40°	50°	60°	70°	80°
η_{T_1}	0.007	0.009	0.016	0.03	0.056	0.12	0.21	0.35	0.44
η_{n_1}	0.010	0.02	0.03	0.06	0.02	0.14	0.21	0.33	0.42
η_{T_2}	0.031	0.056	0.11	0.18	0.25	0.31	0.40	0.49	0.58
η_{n_2}	0.059	0.09	0.15	0.21	0.27	0.33	0.42	0.48	0.57
η_m	0.2	0.25	0.31	0.36	0.41	0.47	0.58	0.67	0.65

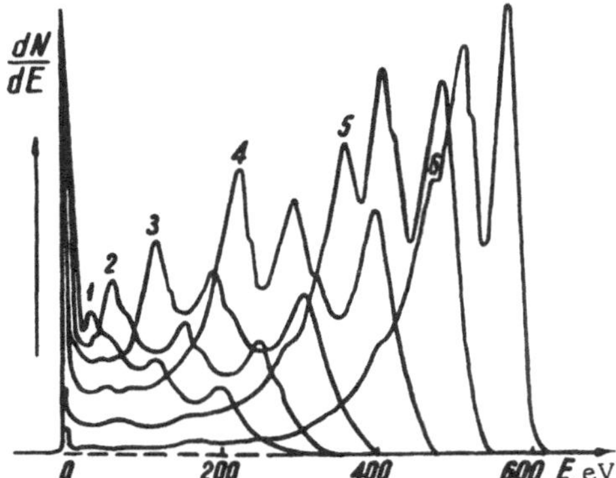

Fig. 4. Series of oscillograms representing the energy distribution of secondary ions obtained for different angles Φ (Θ = 50°): 1) 30°, 2) 40°, 3) 50°, 4) 60°, 5) 70°, 6) 80° in the case of Rb^+ on Mo (100).

dency to deviate from $\eta_{T_1}(\beta)$ and $\eta_{T_2}(\beta)$ in the sense of lower η values over the range $\beta < 40°$. The $\eta_m(\beta)$ curve lies above $\eta_{n_2}(\beta)$ and approaches $\eta_{n_2}(\beta)$ with falling values of the scattering angle β.

Thus the secondary emission current incorporates ions which have suffered many more collisions with target atoms than two. Secondary ions which have experienced one- and two-fold collisions appear as peaks in the spectrum.

The character of the energy distribution of the group of ions with a maximum energy of E_m and the change in this with increasing angle of incidence and primary-ion energy are in no way contradictory to the proposition that these ions arise as a result of more than two collisions. If we consider that after each collision an ion is deflected through an angle β (the same for each collision), then the formula for the energy retained by the ions as a result of multiple collisions at an angle β may be written as follows:

$$E_m = \frac{E_0(\mu - 1)^{2n}}{(\cos\beta' \mp \sqrt{\mu^2 - \sin^2\beta'})^{2n}} \qquad (3)$$

where n is the number of repeated collisions. We see from formula (3) that, with increasing number of collisions n, the value of E_m tends toward E_0. However, with increasing number of collisions the intensity of the multiply-scattered particles with the same or similar scattering angles falls sharply. The faster growth of the values of η_{n_2} and η_m as compared with η_{n_1} in the range $E_0 < 300$ eV, and the displacement of the points at which $\eta_{n_2}(E_0)$, $\eta_m(E_0)$ deviate from the linear law in the direction of greater E_0, may be readily explained as being due to the effect of the binding energy of the target atoms, allowing for the multiplicity of the collisions of the incident ion with the atoms in the target.

References

1. A. A. Aliev and U. A. Arifov, Dokl. Akad. Nauk SSSR, 172(1):65 (1967).
2. U. A. Arifov and A. A. Aliev, Trans. of the Eighth International Conference on Phenomena in Ionized Gases, Vienna (1967).
3. U. A. Arifov and A. A. Aliev, Dokl. Akad. Nauk Uzbek SSR, No. 10, p. 37 (1967).
4. A. A. Aliev and U. A. Arifov, Zh. Éksp. Teor. Fiz., Vol. 54, No. 2 (1968).

5. U. A. Arifov and A. A. Aliev, Dokl. Akad. Nauk SSSR, Vol. 179, No. 5 (1968).
6. É. S. Parilis and N. Yu. Turaev, Dokl. Akad. Nauk Uzbek SSR, No. 12, p. 16 (1964).
7. U. A. Arifov, A. A. Aliev, and A. Kh. Ayukhanov, Izv. Akad. Nauk Uzbek SSR, Ser. Fiz.-Mat. Nauk, No. 4, p. 20 (1964).
8. V. I. Veksler, Fiz. Tverd. Tela, 8:2229 (1964).
9. U. A. Arifov, D. D. Gruich, and L. Kh. Chastukhina, Izv. Akad. Nauk SSSR, Ser. Fiz., 28:1402 (1964).
10. V. M. Kivilis, É. S. Parilis, and N. Yu. Turaev, Dokl. Akad. Nauk Uzbek SSR, 173(4):805 (1967).
11. V. E. Yurasova and D. S. Karpuzov, Fiz. Tverd. Tela, 9:2508 (1967).

EFFECT OF THE STATE OF THE SUBSTRATE ON THE SECONDARY EMISSION OF DIELECTRIC FILMS SUBJECTED TO ION BOMBARDMENT

U. A. Arifov, R. R. Rakhimov, and S. Gaipov

The potential emission of electrons on bombarding alkali halide films with inert-gas ions was mentioned earlier [1]. The coefficient of potential emission γ_{π} was two or three times greater than in the case of refractory metals. The value of γ_{π} increased up to thicknesses of tens of monomolecular layers, i.e., the potential emission of electrons had a depth (three-dimensional) character. However, the mechanism of potential electron emission from a dielectric film deposited on a metal has never yet been elucidated. It would appear likely that the state of the substrate surface, the character of the boundary layer, and also the properties of the actual film should affect the value of γ_{π}.

In this article we shall consider the effect of the state of a molybdenum substrate on the electron and positive- and negative-ion emission arising from NaCl films under bombardment by Ar^+ ions with energies of 50, 100, and 1000 eV.

The experiments were carried out in a vacuum apparatus, the main components of which were described earlier [1]. The arrangement of the receiving part enabled us to study the electron emission separately from the negative ions, the discrimination being effected by means of crossed electric and magnetic fields.

Using the oscillograph double-modulation method [2], we simultaneously measured the coefficients of electron (γ), negative-ion (K_-), and positive-ion (K_+) emission with increasing thickness of the same salt film on a reasonably clean target (substrate) surface. As substrate we used polycrystalline molybdenum in the form of a foil (18 × 7 × 0.02 mm). The substrate was degassed by heating for several hours with the temperature rising gradually to 2200°K. The purity of the molybdenum surface was checked during the degassing process by reference to the value of the potential electron-emission coefficient. Thus after heating 15 h at 2200°K the value of the γ_{π} coefficient obtained for Mo on bombarding with 100-eV argon ions rose as far as 10 to 11%. Experiments were only carried out with the film after achieving a substrate surface of almost atomic purity [3]. In order to preserve a reasonably clean state of the surface, before every deposition of film material the substrate was subjected to brief high-temperature heating. The state of the substrate surface was also varied by oxidizing it at 900 to 1000°K in an oxygen atmosphere at a pressure of 10^{-3} mm Hg for one hour, after which there was a fall in the coefficient of potential electron emission under bombardment with 100-eV Ar^+ ions (between 1 and 2%). The oxide film so formed remained stable on heating to 1200°K, but heating for a few tens of minutes at 2000 to 2200°K removed the oxide film, and γ_π was restored to the value corresponding to the clean state.

In order to determine the thickness of the deposited film, we used the method of amperometric titration, described in another paper [4]. The deposited salt film was dissolved in a small amount of distilled water; by determining the conductivity of this solution, the thickness of the deposited film could be calculated. By using twice-distilled water for dissolving the salt and an ÉO-7 oscillograph for measuring the conductivity of the salt solution, we succeeded in increasing the sensitivity of the method by approximately an order of magnitude as compared with the earlier experiments [4]; it amounted to $2 \cdot 10^{-7}$ g/cm^2, or some 3 to 4 monolayers. The thickness of a specific film was estimated by reference to the time of deposition at a known rate of growth of the layer, the "origin" being taken as the instant at which the substrate heating was disconnected.

In order to determine the laws of potential and kinetic emission separately, measurements were made with three Ar^+ ion energies: 50, 100, and 1000 eV. According to an earlier treatment [3], for energies of up to 150 or 200 eV the coefficient of kinetic emission is quite small for alkali halide films. Hence for energies of 50 to 100 eV the electron emission under Ar^+ bombardment should in the present case be due to the potential mechanism. At an energy of 1000 eV the main part of the electron emission will arise from kinetically emitted electrons.

Figure 1a, b, and c presents the coefficients γ and K_+ in relation to the thickness of an NaCl film deposited on clean and oxidized molybdenum surfaces on irradiating with Ar^+ ions at energies of 50, 100, and 1000 eV respectively. In order to ensure that the film should be deposited on a reasonably clean surface, the latter was briefly heated at 2200°K every time; in the case of the oxidized molybdenum film the temperature employed was no greater than 1200°K, in order to prevent the deterioration of the oxide film on the substrate surface. For small film thicknesses the coefficient γ_π behaved in different ways for the clean and oxidized substrates (Fig. 1a, b). This was evidently because the contribution of the potential emission from the substrate to the total emission differed for a specified film thickness in the two cases.

The decreasing difference between the ordinates of the two curves with increasing film thickness is probably due to the increasing absorption of the electrons emitted from the substrate as the film becomes thicker. In the saturation region, the contribution to the emission from the substrate becomes insignificant, so that the $\gamma_\pi = \gamma_\pi(d)$ curves are the same (within the limits of experimental accuracy) for films deposited on either clean or oxidized substrate surfaces. This indicates that the potential emission is entirely due to the properties of the film itself after the latter has reached the thickness corresponding to the saturation of the $\gamma_\pi = \gamma_\pi(d)$ curve.

We see from Fig. 1c that, even when kinetic emission predominates over potential, the $\gamma = \gamma(d)$ curves are the same for films deposited on clean and oxidized surfaces. The values of γ for any thickness are independent of the state of the substrate. For $E_0 = 1000$ eV the saturation of the $\gamma = \gamma(d)$ curve occurs at a thickness of 50 to 60 monomolecular layers as a result of the comparatively deep penetration of the primary ions into the film, in agreement with earlier data [5].

The positive and negative ion emission is due to the cathodic sputtering of the film material. Comparison of curves 2 and 2′ shows that the negative ion emission depends very strongly on the state of the substrate for small film thicknesses. As the film thickness on a clean substrate increases, the $K_- = K_-(d)$ relationship reaches a maximum at a thickness of about 1 to 2 monolayers, the value of K_- at the maximum being considerably greater than γ for the thickness in question. On further increasing the film thickness, the $K_- = K_-(d)$ curve falls sharply, and a region of saturation ensues. In the case of an oxidized surface, However,

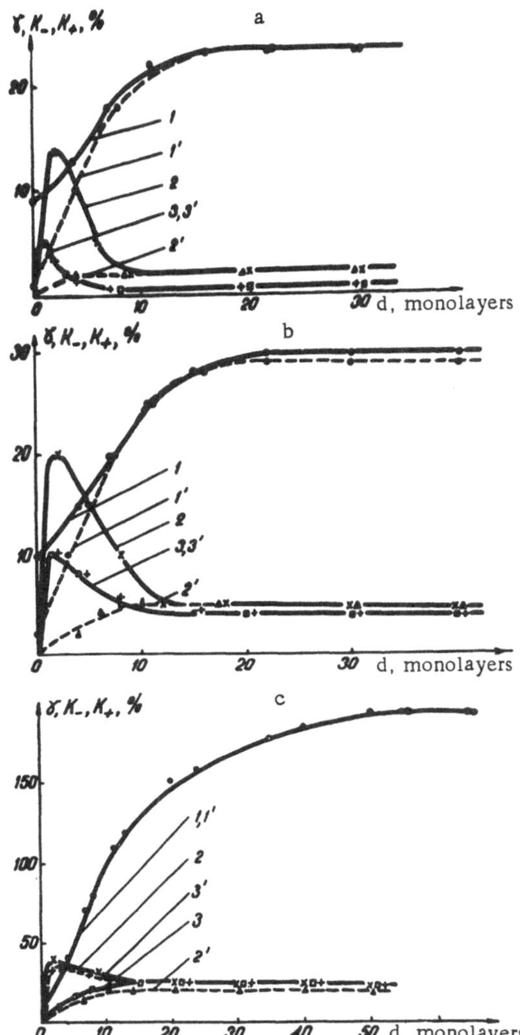

Fig. 1. Dependence of the coefficients γ (1, 1′), K_- (2, 2′), and K_+ (3, 3′) on the thickness of an NaCl film on a Mo substrate bombarded with 50-eV Ar^+ ions (a), 100-eV ions (b), 1000-eV ions (c). Curves 1, 2, and 3 are for a clean and 1′, 2′, 3′ for an oxidized substrate.

K_- rises monotonically up to the saturation value; in the saturation region the substrate has no effect on the value of K_-.

The curves relating the positive-ion emission coefficient to the film thickness for energies of 50 and 100 eV also reach a maximum for a thickness of the order of 1 to 2 monolayers. For greater thicknesses the coefficient K_+ is small and independent of film thickness. It is an interesting fact that for energies of 50 and 100 eV the coefficient of positive ion emission is independent of the state of the substrate surface over the whole range of thicknesses, the $K_+ = K_+(d)$ curves coinciding with each other. For $E_0 = 1000$ eV the $K_+ = K_+(d)$ curve for a film deposited on an oxidized surface reaches a maximum for a thickness of the order of 2 to 3 monolayers, whereas the $K_+ = K_+(d)$ relationship for a film on a clean substrate has no corresponding maximum and rises monotonically to its saturation value.

Thus the coefficient of potential electron emission from NaCl films on Mo in the saturation region is independent of the state of the substrate, which indicates that the emission of the electrons is mainly determined by the properties of the film itself. However, if we take the energy structure of an ideal NaCl crystal and apply it to the NaCl–Mo system, then on bombarding with Ar^+ ions of low energy the potential emission will be negligible owing to the great width of the forbidden band. Experiment shows that the value of γ_π for this system is greater than for the pure substrate, by a factor of two or three times when the film thickness is of the order of tens of layers. We may therefore suppose that the observed rise in the potential emission of the NaCl–Mo system is due to emission associated with interaction between the incident ions and impurity centers and defects in the film.

It was found that, on increasing the energy of the primary ions (from 50 to 100 eV), the coefficient of potential emission also increased (from 23 to 30%). In addition to this, the film thickness at which saturation of the $\gamma_\pi = \gamma_\pi(d)$ curve occurred for an energy of 100 eV was greater than the thickness corresponding to saturation for an energy of 50 eV. This indicates that, with increasing primary-ion energy, interaction takes place between the ions and impurity centers and defects over a long range, owing to the greater depth of penetration of the ions.

In order to secure a better understanding of the effect of impurities on γ and the depth (three-dimensional) character of the potential emission, experiments will have to be carried out with ion and atom beams on films containing impurities of different types. In contrast to the case of ion-electron emission, for small thicknesses the negative ion emission is determined by the state of the substrate; the value of K_- is evidently affected by the conditions of electron exchange between the deposited particles and the substrate, i.e., the effective work function of the surface is of prime importance. This agrees with earlier data [6], according to which the reduction in the effective work function of the surface arising from the adsorption of Cs atoms led to a substantial rise in the negative-ion yield.

References

1. U. A. Arifov, S. Gaipov, M. Ikramova, and R. R. Rakhimov, Dokl. Akad. Nauk Uzbek SSR, 19:11 (1965).
2. U. A. Arifov, A. Kh. Ayukhanov, and S. V. Starodubtsev, Zh. Éksp. Teor. Fiz., 26:714 (1954).
3. H. D. Hagstrum, Phys. Rev., 104(3):672 (1956).
4. M. V. Gomoyunova, Fiz. Tverd. Tela, 1:329 (1959).
5. L. P. Moroz, Author's abstract of Candidate's Dissertation, Physicotechnical Institute, Tashkent (1965).
6. A. Kh. Ayukhanov and M. K. Abdullaeva, Izv. Akad. Nauk SSSR, Ser. Fiz., 30(12):2000 (1966).

ADHESION (BINDING) OF ADSORBED ATOMS IN THE ION BOMBARDMENT OF CLEAN SURFACES

T. D. Radzhabov

The sorption of positive gas ions by a solid differs from ordinary physical adsorption and chemisorption mainly by virtue of the fact that the ions are electrically charged and possess an energy far higher than the thermal energy. The first of these factors is not particularly important for studying sorption processes, since the ion is neutralized before falling on the solid surface [1]. Far more important is the second property, in view of the fact that ions penetrate well into the solid by virtue of their high energy and suffer absorption in the material. In this process, atoms may be freed from the solid target, i.e., the target material is sputtered, and lattice defects are formed in the sites so freed. Some of the gas atoms captured in the solid may become firmly attached to the lattice, while others may migrate from place to place. Subsequently the migrating gas atoms may return to the vacuum after a relatively long period. It is extremely likely that the activation energy will differ from that of normal diffusion, as it may depend on the point of entry of the gas atom.

In giving detailed consideration to the energetics of the adsorption and desorption processes, we may well start by studying the binding, or adhesive, forces acting on the ions falling into the solid. There are various methods for studying the heats and mechanism of desorption processes (field-emission of ions, the use of an adsorption spectrometer, the graphical method, etc.) using high-sensitivity apparatus and devices.

The sorption of ions was studied and the heats of adsorption determined in a number of earlier investigations [2-4] under conditions of very high vacuum, using a monochromatic ion beam and a mass-spectrometer of high resolving power. Using static or dynamic systems for evacuating the working chamber, qualitative and quantitative results were obtained for the sorption of positive gas ions. In order to determine the heats of adsorption, the changes in the partial pressure of the gas under consideration were plotted in relation to the temperature of desorption. The desorption kinetics characterize a thermal process occurring within the lattice; thus by following the theory of desorption (equations of the first order) we may obtain the activation energy from the characteristic temperatures at which peaks of desorption appear. De Boer [5] derived the principal equation for the freeing of sorbed atoms in the form

$$\frac{d\sigma}{dt} = -\sigma\nu_1 \exp\left(\frac{E_d}{RT}\right)\ldots \qquad (1)$$

where σ is the concentration of adsorbed atoms (particles/cm^2),
ν_1 is a constant constituting the vibration frequency of the adsorbed atoms along a coordinate normal to the surface,
$\nu_1 = 10^{13}$ sec^{-1},
E_d is the binding energy (kcal/mole),
R is the gas constant,
T is the absolute characteristic temperature of desorption, °K.

Equation (1) contains a major assumption, namely, it starts from the condition of surface adsorption (σ expresses the surface concentration of adsorbed atoms per square centimeter), whereas, in ion bombardment, sorption occurs in any layer of the material. Owing to the absence of adequate data relating to the depth of penetration of ions into matter, we cannot determine the exact extent of the saturated layer. However, on the basis of available results [6], we may consider that the depth of penetration in the case of a metal irradiated with inert-gas ions fluctuates up to the extent of several atomic layers. Thus the de Boer equation (1) may properly be employed for an approximate calculation of the heats of adsorption; the accuracy of the determination will in this case be about ± 3 kcal/mole.

For the case of gas on glass, no published penetration depths are available. James and Carter [7] studied the desorption of inert gases from glass by the secondary-bombardment method, and concluded that the gas was largely concentrated in the surface layers, apparently as a consequence of the structure of the glass. Desorption also obeyed Eq. (1) in this case.

On comparing the interatomic distances in glass with the diameters of the incident ions, we see that the former are very small (for example, in SiO, silicon glass, the interatomic distances are only 1.6 Å). Since the diameters of the inert-gas ions (apart from helium) have appreciably larger dimensions than the interaction distances of glass, the ions are clearly unable to penetrate between the atoms, and the thermal-desorption curves will have a sharp, single-peak characteristic, i.e., desorption will occur only at the surface or in defective places, surface cracks, pores, etc. Cobic, Carter, and Leck [8] indicate that the main proportion of the gas (about 95%) is desorbed from the glass at temperatures of around 300 to 350°C, which makes the heat of adsorption 30 to 35 kcal/mole. These results are also supported by data relating to the diffusion coefficients of noble gases in various glasses. The mechanism of the ionic pumping of gases by glass "in cracks" is supported by the fact that the maximum amount of gas capable of being pumped out by the glass is always smaller than the number of particles required to cover the surface with a monolayer. Schwarz [9] even indicated that a glass surface acted as an ion trap.

When studying the thermal desorption of inert gases (absorbed on ion bombardment) from a metal such as tungsten, Kornelsen [10] observed six characteristic peaks on heating the target to 2400°K, and divided these into two groups α and β; the α group included gas atoms penetrating several lattice constants into the target, while the β group included atoms which had penetrated more deeply than a few lattice constants into the material, up to 0.1 μ.

The temperatures associated with the peaks of the α group are not strictly dependent on the energy of the ions, while the peaks of the β group become stronger with increasing ion energy.

It should nevertheless be noted that Colligon and Leck [2], when studying the thermal desorption of inert gases from W, Pt, and Mo, observed only single-peak and two-peak characteristics. The desorption curves were taken as one- or two-peak according to the form of the ion and the crystal lattice of the target. A detailed analysis of the two forms showed that the temperatures characteristic of the first (low-temperature) peak should be associated with the α group of the Kornelsen classification, and those characteristic of the second (high-temperature)

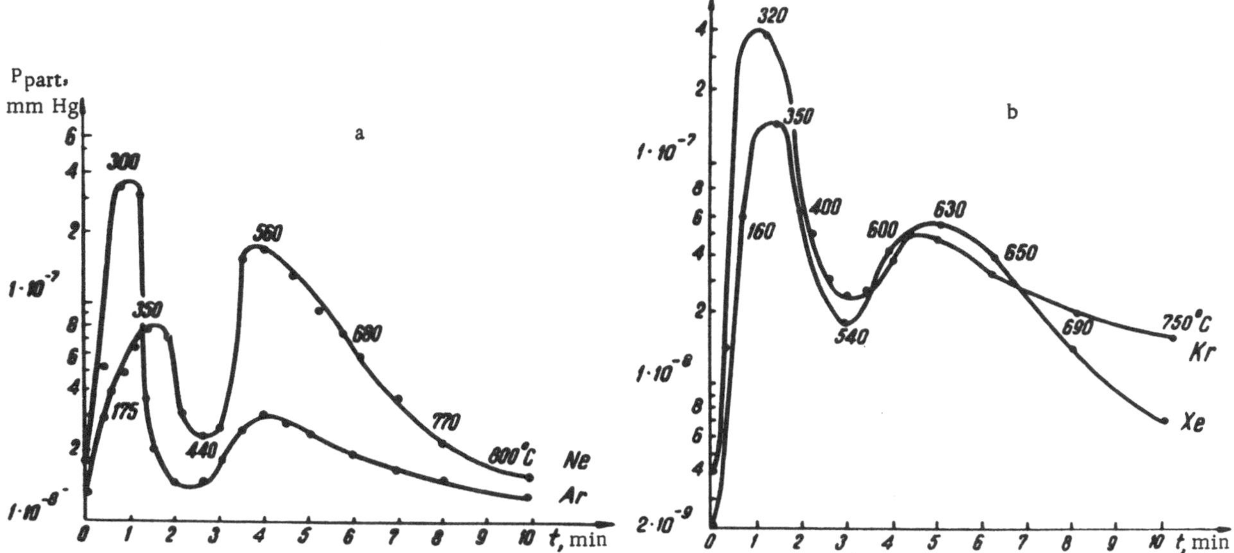

Fig. 1. Desorption of argon, neon (a), krypton, and xenon (b) from freshly deposited titanium film.

peak with the β group. The curves representing the thermal desorption of Ne and Kr from Pt and Mo had a single-peak character, with desorption energies of 36 and 60 kcal/mole respectively.

Let us give the desorption energies for a tungsten target determined by Colligon and Leck:

Gas peak	Ne	Ar	Kr
First	50	33	43
Second	80	70	—

The two-peak character of the desorption process is (according to the authors) most probably due to the existence of two sorption mechanisms in the crystal lattice of the target (for example, inclusion and substitution), with different binding energies. The existence of several intermediate peaks in each group, as obtained by Kornelsen, may be associated with surface migration and diffusion.

In our own investigations [11-13], the spectrum representing the thermal desorption of inert gases from titanium was studied up to a temperature of 900 to 1200°K; naturally, fewer desorption peaks were obtained in this case. In the case in question, the characteristic temperatures were shifted in the low-temperature direction. The range of temperatures selected and the use of targets with different grain sizes (and also single crystals) enabled us to determine the character of the low-temperature peaks quite clearly. On the basis of our experiments, we established that the low-energy peaks, with a heat of desorption of 30 to 35 kcal/mole, related to ions sorbed along the grain boundaries and in crystal-lattice defects, and the high-temperature peaks, with a heat of desorption of 50 to 55 kcal/mole, to ions sorbed by direct incorporation in the crystal lattice. Our characteristic curves of thermal desorption obtained on irradiating titanium with inert-gas ions are presented in Figs. 1 and 2, and the curves representing desorption from a single crystal appear in another paper [12]. These graphs illustrate the existence of two binding mechanisms (two-peak curves of thermal desorption). As the dimensions of the gas ion diminish, the possibility of the simultaneous occurrence of both mechanisms does likewise [12]. In fact, in our experiments with helium ions (which have a diameter smaller than the interatomic distance in the titanium lattice), the desorption curves

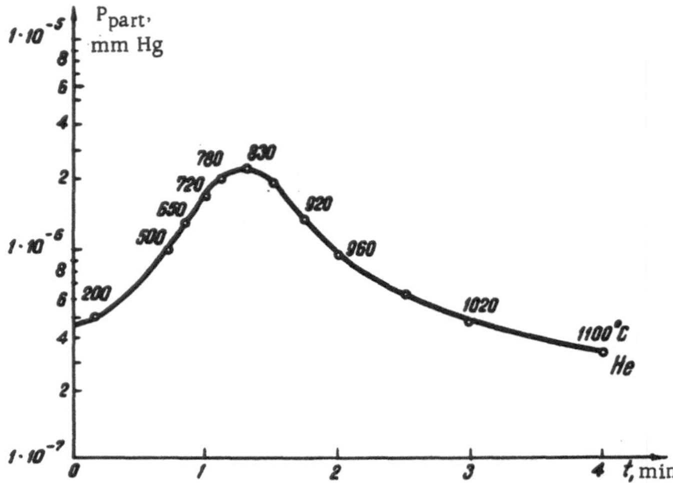

Fig. 2. Desorption of helium from a freshly deposited titanium film.

have an obvious single-peaked character. Our assumption as to the sorption of the ions along grain boundaries and in lattice defects is borne out by the experiments with the titanium single crystal, which retained a single, high-temperature peak, corresponding to ions which had penetrated directly into the crystal lattice of the target.

We obtained the following values for the heats of desorption corresponding to titanium films (kcal/mole):

Film	Peak	He	Ne	Ar	Kr	Xe
Freshly deposited	First	52,6	31	28.5	29,6	30.9
	Second	—	42	41	44,4	45.0
Regenerated	First	55	30.9	30,6	30.1	36.6
	Second	—	42.5	42	45.0	46.0

The changes in the heats of desorption of argon for various titanium structures are as follows (kcal/mole):

Gas peak for Ar	Sputtered Ti film	Compact Ti	Coarse-crystalline Ti	Single crystal Ti
First	28.5	34	38.6	47
Second	41	46	45	–

Unfortunately, we were unable to record the amount of gas evolved at temperatures close to the melting point, at which we should have expected the evolution of gas atoms sorbed by direct substitution in the crystal lattice. However, there was a considerable increase in the desorption of the gas on approaching the onset of melting in the target material.

There are several points of view regarding the mechanism of the sorption of ions, and these may be categorized in the following manner [14]:

1) The hypothesis of "ion shock" (or impact), in which the gas ions penetrate to a greater or lesser degree into the target material under the influence of an electric field and then partly emerge into the gas space. In this case the structure of the gas-evolving material is of particular importance.

2) The idea that sorption occurs, not as a result of the acceleration of the flow of ions, but under the influence of Langmuir adsorption forces; in this case the sorption is predominantly a surface effect.

3) The hypothesis of "centers," according to which the binding of the gas particles to prepared parts of the target surface is independent of the character of the ions.

On the basis of the experiments carried out, we may now refine our model for the sorption of inert-gas ions by pure metals. An analysis of the results obtained tends to favor the first hypothesis, the one associated with the depth of penetration and the structure of the material. The second model is chiefly applicable to such chemically-active gases as nitrogen, which are able to dissociate; the role of ionization in this case reduces to the production of excited (activated) gas particles. The hypothesis of "centers" probably relates more to the material of the target itself, and thus predominates in the case of such materials as silicagel and zeolite, being less applicable in the case of metals.

According to the "ion shock" theory, in the ion bombardment of metal surfaces the gas atoms may be sorbed on the target surface, in crystal-lattice defects, pores, and intercrystallite zones, directly in the crystal lattice of the target material, or by replacing existing lattice atoms. Corresponding to these several groups, the energy of desorption may fluctuate from a minimum of 30 kcal/mole to a maximum of 90 to 100 kcal/mole.

Subsequent investigations into the thermal desorption of gases sorbed by ion bombardment at temperatures close to the melting point, and during the actual melting of the target material, should enable us to verify the validity of this mechanism.

References

1. H. D. Hagstrum, Phys. Rev., 104:317 (1956).
2. J. S. Colligon and J. H. Leck, Trans. Eighth Nat. Vac. Symposium, Washington, October 1 (1961), p. 275.
3. R. B. Burtt, J. S. Colligon, and J. H. Leck, Brit. J. Appl. Phys., 12(8):396 (1961).
4. J. H. Carmichael and E. A. Trendelenburg, J. Appl. Phys., 29:1570 (1958).
5. J. de Boer, Dynamic Character of Adsorption [Russian translation], IL, Moscow (1962).
6. J. S. Colligon, Vacuum, II(5/6):272 (1961).
7. L. K. James and G. Carter, Brit. J. Appl. Phys., 13:1 (1962).
8. B. Cobic, G. Carter, and J. H. Leck, Brit. J. Appl. Phys., 12:288 (1961).
9. H. Schwartz, Z. Phys., 117:23 (1940); 122:437 (1944).
10. E. V. Kornelsen, Vak.-Techn., Vol. I (1964).
11. G. F. Ivanovskii and T. D. Radzhabov, Zh. Tekhn. Fiz., 35(7): 1313 (1965).
12. G. F. Ivanovskii, T. D. Radzhabov, and T. N. Zagorskaya, Zh. Tekh. Fiz., 36(8): 1469 (1966).
13. T. D. Radzhabov [Radjabov] and G. F. Ivanovskii [Ivanovsky], Trans. Third Inter. Vac. Congress, Stuttgart, Vol. 2 (1966), p. 1.
14. G. Strotzer, Zeit. f. Ang. Phys., 10(5):207 (1958).

SORPTION OF GASES IN THE ION BOMBARDMENT OF GLASS SURFACES

T. D. Radzhabov

The development of very high vacuum techniques and solid-state physics demands a more careful investigation into the nature and mechanism of the interaction of positive gas ions with glass surfaces. In one of the very earliest investigations of Campbell [1] relating to the association between the "blue discharge" and the evacuation of various gases, it was asserted that chemically-active gases such as hydrogen, carbon monoxide, and nitrogen were sorbed more rapidly than inert gases (such as argon). Since the mechanism of the sorption of argon was incapable of being explained by chemical reactions or dissociation, Campbell suggested that absorption in the discharge depended on the nature of the gas and the purity of the glass walls of the tube rather than constituting a purely chemical phenomenon.

Andrews, Hutsteiner, and Dushman [2] studied the evacuation of nitrogen and argon at a pressure of 1 μ Hg in the absence of the blue glow, and came to the conclusion that the rate of evacuation varied almost linearly with the gas pressure and electron current (while this remained under 1 mA).

Later [3] Campbell studied the sorption of N_2, CO_2, O_2, H_2, and Ar in a sealed tube of the triode type. On the basis of this and other investigations, it was established that, in the presence of a discharge, the gases were all evacuated to different extents; the mechanism of the reaction depended on the nature of the gas and the conditions of the discharge, and also on the state of the walls of the discharge tube; the sorbed gas was repeatedly evolved and absorbed again by the walls.

As the pressure falls, so does the rate of ion formation and hence the rate of evacuation (for a constant electron current). Von Meyern [4] sought to increase the number of ions formed at reduced pressure by increasing their distance of travel, making the anode in the form of a ring, while the discharge tube was placed in a solenoid, as a result of which the electrons moved along a spiral trajectory. It was found that the total amount of absorbed gas never exceeded the amount required to form a monolayer on the walls. Under these conditions equilibrium was achieved. In studying the absorption of helium, nitrogen, and argon, Von Meyern established that argon was evacuated in accordance with the law

$$-\frac{dP}{dt} = U(P - P_e) \quad (1)$$

where P is the instantaneous pressure,

P_e is the pressure established at equilibrium,

U is a constant depending on the magnetic field and anode voltage.

TABLE 1.

Author	Equation
Alpert [5, 6]	$-dP/dt=(P-P_U)S/V$
Young [7]	$-dP/dt=a+bP+cP^2$
Carmichael and Varnerin [8]	$-dP/dt=AP-f(t)$
Robinson and Berz [9]	$-dP/dt=\alpha \exp(-\beta t)+\gamma \exp(-\delta t)$

Table 1 presents similar equations of various authors in simplified form with unified notation, in which S, V, P_U, a, b, c, A, α, P, γ and δ are constants, $f(t)$ is a function of time, and P is the pressure of the gas.

The most suitable equation is that of Robinson and Berz.

Bayard and Alpert [5, 6] described the operation of a very high-vacuum ionization manometer and studied its evacuation properties. It was found that, if the pressure were held at a level of 10^{-4} mm Hg and gas were admitted continuously, the pumping action ceased after an hour. Saturation was reached with the formation of approximately one monolayer. If the pressure is equal to 10^{-9} mm Hg, the manometer will pump for no less than three months without any sign of saturation. Assuming that the pumping action was simply associated with the removal of gas ions, Alpert demonstrated excellent agreement between theory and experiment.

Bloomer and Haine [10] used the Alpert method in a detailed study of the evacuation of residual gases with ionization manometers. In order to verify the proposition that the evacuation was associated with the formation of ions, the authors varied the electron emission current and related the pumping speed to the rate of ion formation. The pumping curve consisted of two sections: In one the relationship between the pumping speed and the emission current was linear; in the other the linearity was preserved for an emission current of under 50 μA, while for higher currents the pumping speed was independent of the emission current. Bloomer and Haine concluded that the evacuation was a consequence of something more than simply the removal of ions. Excited molecules were of considerable importance in the pumping process. No other explanation could satisfy the observations.

Theories explaining the mechanism of ionic pumping by glass surfaces may be separated into two groups:

1) Carleton, Bills, Young, and several other research workers considered the pumping as a surface phenomenon taking place at specific "trapping points" and producing a spontaneous secondary emission of earlier-trapped gas;

2) Varnerin and Carmichael suggested the penetration of positive ions into the solid surface. The secondary emission of the gas greatly complicates this mechanism, but it may be monitored in the course of diffusion.

Varnerin and Carmichael [11] studied the effect of metallic films, deposited on the inner surfaces of Bayard–Alpert manometers, on the evacuation of helium. Figure 1 shows the pumping curves, the instantaneous pressure being normalized by expressing it as a fraction of the initial pressure. Curves 1 to 4 correspond to different thicknesses of the metallic film deposited on the walls of the manometer (1 represents the minimum and 4 the maximum). It was found that, on increasing the deposit of metallic tungsten on the glass walls (films up to 50 Å thick were studied), the rate of evacuation by the walls increased; the rate of gas desorption was proportional to the amount of absorbed gas and the time of measuring the rate of desorption after pumping. The authors gave no explanation for the mechanism by which the gas

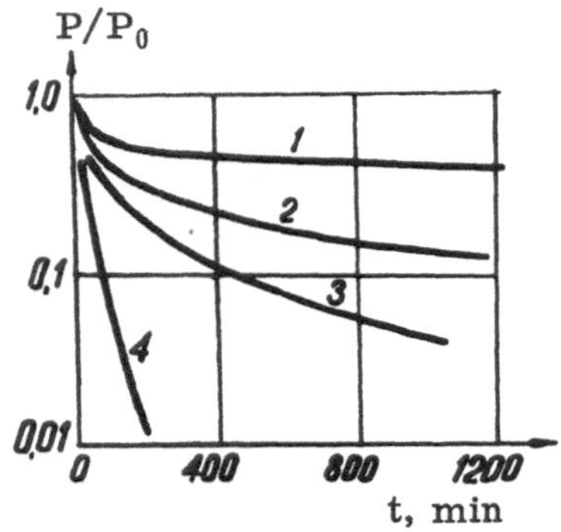

Fig. 1. Pumping curves of helium in a sealed Bayard–Alpert manometer. Curves 1, 2, 3, and 4 correspond to a gradual increase in the thickness of the film.

was evolved into the surrounding volume. The rate of sorption constituted the sum of the rates of adsorption and desorption. The reduction in the rate of evacuation of helium observed at liquid-nitrogen temperature evidently arose from the spontaneous evolution of the gas.

The prolonged reverse evolution of the helium absorbed in ionic pumping was inversely proportional to the time of measurement. The same relationships were obtained for argon, neon, and krypton absorbed by molybdenum and nickel targets [12]. On this basis a formula was derived for the rate of desorption

$$r = k\frac{N_t}{t} \quad (1 < t < 250 \text{ min}) \tag{2}$$

where N_t is the amount of gas evacuated,
t is the time,
k is a constant depending on the energy of the ion.

According to Carmichael and Knoll [13], for an ion energy of about 200 eV the values of k for the inert gases were the following:

Gas	Nickel × 10^2	Molybdenum × 10^2
He	4.4 ±1.8	1.3±0.05
Ne	5.2 ±4.6	4.0±0.3
Ar	0.57±0.07	15±5
Kr	0.32±0.16	14±1

In none of the foregoing investigations was the reverse gas pressure measured as a function of time for periods shorter than one minute; shorter intervals were studied by Fox and Knoll [14]. For periods of time down to 0.03 min the reverse evolution was proportional to $1/t_r$ (t_r = time of reverse evolution). The effect of anode temperature on the pumping speed and reverse gas pressure was also studied [14]. The experiments showed that, on reducing the temperature of the anode walls, not only did the pumping time constant diminish but the amount of gas evacuated also increased. The reverse evolution of the gas diminished with falling anode temperature. The trapping coefficient was also measured [14] as a function of the surface temperature for an incident ion energy of about 200 eV:

Wall temperature, °K	330	190	77
Mean value of τ, min	2.5	1.1	0.7
Reverse emission constant, k	$6.5 \cdot 10^{-2}$	$5.4 \cdot 10^{-2}$	10^{-3}

For an anode temperature equal to the temperature of liquid nitrogen, the rate of reverse gas evolution could not be measured.

The trapping efficiency for the capture of inert-gas ions (Ne, Ar, Kr, and Xe) by Pyrex glass increases rapidly from 0.1 at 25 eV to 0.9 at 250 eV for Ar, Kr, and Xe, while for Ne the efficiency varies only slightly and remains between 0.1 and 0.2 [15].

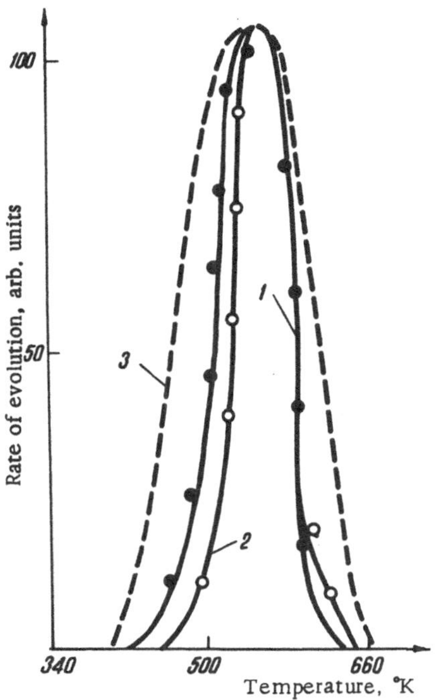

Fig. 2. Desorption of helium from glass surfaces contaminated with tungsten films: 1) theoretical (Q = 34 kcal/mole); 2) experimental, for a thin layer of tungsten; 3) experimental, for a thick layer of tungsten.

In another investigation [16], Grant and Carter made a detailed study of the ionic pumping and thermal desorption of He from Pyrex glass. It was found that, within the limits of experimental sensitivity, well-cleaned glass absorbed no He ions; appreciable helium absorption only appeared after deposit of thin tungsten films on the glass surface of the diode. The trapping efficiency reached 0.1 at 250 eV. This agreed closely with the results of Corkhill and Carter [17] and Erents and Carter [18] obtained in experiments with an ion beam on a tungsten strip; however, it disagreed with the results of Cobic, Carter, and Leck [19], according to which the trapping efficiency equalled 0.87 at an energy of 250 eV and was independent of the presence of any appreciable metallic layers. A trapping efficiency of 0.2 for ion energies of about 200 eV in the ionic pumping of helium on clean glass surfaces was indicated by Alpert [20], the corresponding results agreeing closely with the data of Varnerin and Carmichael [11].

The curves representing the desorption of helium from thin and thick tungsten films obtained by Grant and Carter [15] are shown in Fig. 2.

Young [7] studied the pumping of helium in a Bayard–Alpert manometer by a technique differing from that of Varnerin and Carmichael. In this case the experiments were carried out in a hermetized manometer. A movable glass casing was placed on the electrode system, protecting the manometer walls from contamination. Young came to the conclusion that, in order to ensure a minimum helium pumping speed, a slight metallic deposit was essential. The maximum helium pumping speed was 0.004 liter/sec for an emission current of 10 mA, while for nitrogen and air the value was 0.1 liter/sec, the shielding of the glass walls not having any appreciable effect in this case on the pumping. In order to elucidate the influence of bombardment on the pumping speed, Young covered the inner surface of the walls with aquadag; he found that the number of particles falling on the walls was 5 to 11 times greater than on the collector.

Comša and Muša [21] described an ionization pump in which accelerated electrons escaping from a tungsten cathode placed in the center of the tube were directed toward a cylindrical electron collector. The ions so formed were collected on the glass walls. The rates of pumping the residual gases obtained in this tube remained constant over the range 10^{-5} to 10^{-8} mm Hg and were higher than those of Alpert, presumably as a result of the fact that more ions were formed in the volume.

Bills and Carleton [22] studied the pumping of nitrogen and oxygen by means of an Alpert manometer. The results agreed with the observations relating to helium absorption. In contrast to the other research workers, who found a reduction in the pressure inside a closed manometer, Bills and Carleton maintained a constant pressure by continuously introducing fresh portions of gas through a variable leak; they found on using the method of the closed volume it was impossible to conduct a valid analysis of the results, even for a minimum reverse evolution of gas. The cessation of the pumping action in this case may be attributed to the lack of gas in the pump.

A system with a continuous gas flow leaking through the manometer was employed by Leck and Carter as well [23]. The amount of absorbed gas was established, after extinguishing the discharge, by heating the manometer to 400°C; the pressure was determined by reference to a mass spectrometer or ionization manometer. In studying the sorption of argon it was observed that, for any energy of the ions, two different amounts of gas might be collected, this effect depending on the different states of charge on the glass walls. The charge on the walls influenced the secondary emission, so that any estimate of the ion fluxes was inaccurate. The maximum rate of argon pumping was 0.004 liter/sec for an emission of 1.0 mA, and the maximum amount of absorbed gas was approximately 15% of a monolayer covering the glass. Similar experiments were carried out for Kr and Ne, using a mass spectrometer. The pumping speed was approximately the same as for argon. A double-valued sorption effect was also noticed in the case of Kr. The walls of the monometer were not shielded from the metallic deposit. The results obtained in a manometer degassed at 400°C differed in no way (within the limits of experimental error) from the results obtained over many hours of operation. This last assertion would appear doubtful, as a perceptible effect would be expected from the deposit so formed. It should be noted that this work was carried out with heavy inert gases and not helium, so that it in no way contradicts the results of Varnerin and Carmichael.

Cobic, Carter, and Leck [24] varied the electron current from 0.1 to 10 mA, the electron accelerating voltage from 0 to 1000 V, and the tube temperature from −196 to +400°C, and studied the pumping of He, Ar, Kr, and Xe in a Bayard–Alpert manometer. It was found that the initial rate of sorption increased linearly with the emission current at constant grid voltage and was independent of gas pressure (down to 10^{-7} mm Hg); it increased with increasing atomic weight of the gas and fell with increasing temperature, increasing with rising grid voltage and reaching a maximum for heavy noble gases at a voltage of about 500 V. The pumping speed fell with increasing amount of sorbed gas. The maximum sorption probability, i.e., the ratio of the number of sorbed ions to the number of incident ions, depended on the atomic weights of the gases.

For helium the sorption probability reached a maximum, for argon it fell to a minimum, while for Nr, Kr, and Xe it was almost the same in each case. These results imply the existense of sorption points with different binding energies, the occupation of these demanding an activation energy of some 25 to 50 kcal/mole.

Comparative characteristics of the pumping of argon and nitrogen with a Bayard–Alpert manometer were secured by Carter and Cobic [19]. In the theoretical derivations it was assumed that three processes were taking place in the manometer:

1) The arrival of excited gas particles on unoccupied parts of the surface and their capture (with a certain degree of probability);

2) the arrival of excited particles on occupied parts of the surface and the partial transfer of sorbed particles into the gas phase;

3) the evolution of gas sorbed on parts of the surface with a low binding energy.

On the assumption that all the ions formed in the manometer are single-charged and that they bombard the surface with the same energy and in the same direction, the probability of the different processes taking place is very much the same for all parts of the surface. A sharp division may be made into regions with low and high energies respectively. Analysis of the results enables us to relate the rate of gas pumping on specific parts of the surface, characterized by different energies, to the rate of desorption and the density of the gas. The conclusions of Alpert (regarding the constant pumping speed) may be explained by the phenomenon of saturation, although a variable pumping curve is usually obtained (Von Meyern's data constitute an exception). In the case of nitrogen we may clearly interpret the pumping on the basis of the

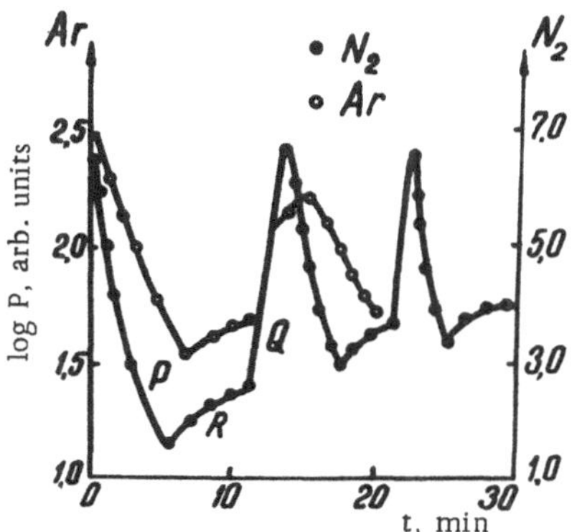

Fig. 3. Repeated evacuation and evolution of Ar and N_2: P) pumping region, R) evolution region, Q) admission of fresh gas.

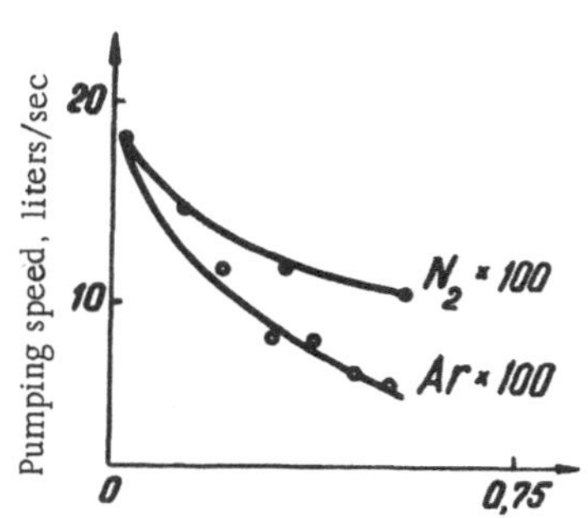

Fig. 4. Pumping speed in relation to the amount of gas evacuated.

Young, Bills, and Carleton mechanism by assuming a limited number of capture points and the gradual filling of these, leading to saturation. Spontaneous evolution clearly constitutes a severe simplification of the problem. In order to explain the sorption of helium in terms of the results of Varnerin and Carmichael [8, 11], we may imagine a reduction in the number of capture points and a time-dependence of the desorption of the absorbed gas.

Carter [26] presented some data relating to the maximum amount of gas evacuated (mm Hg • liters) as a function of the nature of the glass:

	Ne	Ar	Kr	Xe	N_2
Quartz glass	1.7	3.05	1.7	1.26	3.40
Pyrex	0.25	2.12	2.5	1.90	4.0
Lead glass	0.13	1.05	1.7	1.15	3.50

In order to determine the initial pumping speed and the maximum amount of absorbed gas, a method based on the alternate admission and desorption of the absorbed gas was employed. Figure 3 shows a typical curve of the evacuation and evolution of Ar and N_2. Figure 4 shows some characteristic curves relating the pumping speed to the amount of gas evacuated for argon and nitrogen (the measuring tube worked at room temperature and an electron-accelerating voltage of 250 V).

James and Carter [27] studied the desorption of inert gases sorbed by glass surfaces by secondary ion bombardment (combinations Ar → Kr, Kr → Ar). It was found that the release of the absorbed gas in the course of ion bombardment took place mainly as a result of the evolution of gas due to the sputtering of the glass surface, the gas being mainly concentrated in the surface layers and not penetrating further into the material. This may be explained by the particular structure of the glass surface. These results [27] agree closely with the earlier data [28]. Later the same authors [29] calculated the gas-evolution coefficient I_g from the formula

$$I_g = \frac{\text{Number of desorbed atoms A}}{\text{Number of incident ions B}} \frac{1}{\theta} \quad (3)$$

where Θ is the relative degree of occupation of the surface with the gas A.

Thus the pumping of gases in glass discharge tubes is mainly the result of the formation of activated gas particles, arising in collisions with electrons. The ions so formed interact with the walls of the vessel, receive an energy of the order of several hundreds of electron volts, and bombard the low-potential glass. At normal temperature the atoms of the glass envelope vibrate with a thermal energy of a few tens of electron volts; hence the interaction between the bombarding ions and the glass is completely dominant. The forces associated with physical adsorption at room temperature determine the lifetime of adsorbed atoms or molecules on the surface. This time is shorter than one microsecond. It is clear, however, from the majority of the experiments that the particles here under consideration remain for a long time on the surface before starting to evolve. Thus Liverpool showed that particles remained in contact with the glass at 375°K for over an hour. This proves that simple physical adsorption cannot explain the mechanism holding the gases in ionic pumping. As regards the possibility of chemisorption and chemical reactions, these processes certainly have a decisive effect in getter tubes. However, ionic pumping in nongetter tubes cannot be explained by chemisorption, nor can the pumping of inert gases.

It would appear that gas ions with a high kinetic energy, bombarding the glass walls, are able to penetrate between the atoms of the glass lattice. The depth of penetration depends on the energy of the incident ion, the angular distribution, the size of the ion, and the atomic structure of the surface. The phenomenon of desorption by secondary ion bombardment is also explicable, since the bombarding ion may remove the captured particles by opening the lattice for a time sufficient to allow the particles to escape by virtue of thermal energy transfer and local heating. It is quite possible that the sorbed particles are situated on asperities of the surface structure, and that the impact of the incident ion may communicate sufficient energy to release the particle.

We see from the foregoing review that, although a considerable amount of experimental material has now been gathered together, there remains a certain lack of clarity in the mechanism underlying the interaction of positive gas ions with glass surfaces. More detailed study of the physical nature of the sorption of gas ions by glass surfaces will be required, as well as a comparison of existing data with the sorption of ions by metallic surfaces, and the use of more sensitive methods and apparatus. This will enable us to gain a better idea of the nature and mechanism of the sorption of gases by glass, elucidating the general laws governing the sorption of gas ions by solids, and leading to practical recommendations in relation to the construction of electrical vacuum apparatus, experimental physics, and high-vacuum technology.

References

1. N. R. Campbell, Phil. Mag., 41:685 (1921).
2. S. Dushman, Fundamentals of Vacuum Technology [Russian translation], Izd. Mir, Moscow (1964).
3. N. R. Campbell, Phil. Mag., 41:778 (1924).
4. W. Von Meyern, Z. Phys., 84:531 (1933).
5. R. T. Bayard and D. Alpert, Rev. Sci. Inst., 21:571 (1950).
6. D. Alpert, J. Appl. Phys., 24:860 (1953).
7. J. R. Young, J. Appl. Phys., 27(8):926 (1956).
8. L. J. Varnerin and J. H. Carmichael, J. Appl. Phys., 28:913 (1957).
9. N. W. Robinson and F. Berz, Vacuum, 3:48 (1953).
10. R. N. Bloomer and M. E. Haine, Vacuum, 3:128 (1953).
11. L. J. Varnerin and J. H. Carmichael, J. Appl. Phys., 26:782 (1955).
12. J. H. Carmichael and J. S. Knoll, Trans. Fifth Nat. Symp. Vac. Techn., Pergamon Press (1959), p. 18.
13. R. E. Fox and J. S. Knoll, Trans. Seventh Nat. Symp. Vac. Techn. (1960), p. 306.
14. J. H. Carmichael and J. S. Knoll, Westinghouse Research Sci. Paper 6-94436-1-P1 (1958).

15. W. A. Grant and G. Carter, Proc. Seventh Conf. Ionization Phenomena in Gases, Belgrade, Gradevinska Kniga, Belgrade,(August, 1965).
16. W. A. Grant and G. Carter, Vacuum, 16(9):485 (1966).
17. D. P. Corkhill and G. Carter, Proc. Conf. Electromagnetic Isotope Separators, Related Ion Accelerators, and Their Application to Physics, Aerhus, Denmark (1965); Publ. in Nuclear Inst. Methods, 38:192 (1965).
18. K. Erents and G. Carter, Vacuum (in press).
19. B. Cobic, G. Carter, and J. H. Leck, Vacuum, 11:247 (1961).
20. D. Alpert, Handbuch der Physik, 12:609 (1958).
21. G. Comša and G. Muša, J. Sci. Inst., 34:291 (1957).
22. D. G. Bills and N. P. Carleton, J. Appl. Phys., 29:692 (1958).
23. J. H. Leck and G. Carter, First Intern. Vac. Symp., Namur (1958).
24. B. Cobic, G. Carter, and J. H. Leck, Brit. J. Appl. Phys., 12:288 (1961).
25. B. Cobic and G. Carter, Le Vide, 100(7):320 (1962).
26. G. Carter, Vakuum Technik, 2:47 (1963).
27. L. H. James and G. Carter, Brit. J. Appl. Phys., 13:1 (1962).
28. G. Carter and J. H. Leck, Proc. Roy. Soc. A261:303 (1961).
29. L. H. James and G. Carter, Trans. Ninth Nat. Vac. Symp., New York (1963), p. 406.

SECONDARY EMISSION OF THIN FILMS ON PASSING IONS OF CERTAIN ALKALI METALS THROUGH THEM

Sh. A. Ablyaev and N. Kh. Dzhemilev

The excitation of electrons by atomic particles with energies ranging from tens of electron volts to tens of kiloelectron volts in solids constitutes a problem of great scientific and practical interest. Up to the present time, information relating to this kind of excitation has usually been obtained by studying ion–electron emission in reflection; this information has proved inadequate for a complete explanation of the phenomena observed. Further investigations are required.

Valuable information relating to these processes may also be secured by studying secondary-electron emission in transmission (incident beam shot through a thin film). Of great interest in this respect is the energy spectrum of the electrons emitted, as this characterizes the process better from a qualitative point of view that the secondary-emission coefficient, although the latter reveals the probability with which the excited electrons escape.

A study of the energy distribution of truly secondary electrons in transmission indicates that these distributions are represented by curves with maxima. Quantitative data as to the position of the maxima $E_{max}^{!2n}$ and their half width ΔE are in poor mutual agreement [1-4]. The electron energy distribution obtained under the impact of beams of 57-keV Li^{+6n} ions and 5-keV electrons was studied by W. Dietrich and H. Seller [5]. It was found that the electron energy distribution depended on the material; it was broader under bombardment with electrons than under bombardment with ions; KCl had the narrowest distribution.

There are no data relating to the secondary-emission coefficient in transmission at the present time. We accordingly decided to study the secondary-electron emission of thin films on passing alkali metal ions through these.

An analysis of the secondary-electron energy spectrum was carried out at the same time as a measurement of the secondary-emission coefficients by the spherical- and cylindrical-condenser methods. In order to accelerate the measurement of the energy spectrum we used an oscillograph method [6]. The combination of these methods enabled us to combine, in a single piece of apparatus, the high resolving power of the cylindrical condenser, the high transmission of the spherical condenser, and the high-speed recording of the oscillograph method of measuring secondary processes.

The vacuum apparatus consisted of an electron and ion source (source and focusing lenses), a spherical collector, a 127° cylindrical condenser with a 2% resolving power, and an electron multiplier.

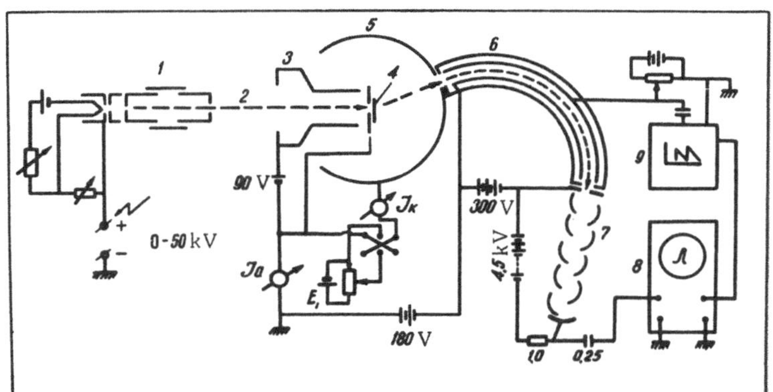

Fig. 1. Fundamental arrangement of the experimental apparatus: 1) ion source with focusing system; 2) primary beam; 3) diaphragm; 4) emitter; 5) spherical collector; 6) 127° cylindrical condenser; 7) electron multiplier; 8) oscillograph; 9) generator.

The source, the receiving section together with the analyzer, and the electrical measuring circuit are shown in Fig. 1. The primary beam 2, accelerated to a specific energy, fell normally on the target surface 4. The majority of the secondary electrons were collected by the spherical collector 5, the potential of this relative to the target being defined by the battery E_1. Some of the secondary electrons passed through the slit of the collector at a small angle to the target surface and fell into the space of the cylindrical analyzer 6, the field of this being varied by means of a sawtooth generator 9.

Electrons with a wide energy distribution passed through the electrostatic analyzer into an electron multiplier 7 having an amplification factor of 10^8. The amplified current was recorded by an oscillograph 8, the horizontal scan of this being synchronized with the sawtooth voltage generator.

Thus the oscillograph screen automatically recorded the energy-distribution curve of the secondary electrons. The values of the secondary-emission coefficients γ and σ were determined as the ratio of the emission current from the surface (or more exactly, the current I_k in the collector) to the beam current I_b of the primary beam, allowing for the transmission factor of the grid, by using the formula

$$\sigma \text{ or } \gamma = \frac{I_k}{I_b a} - I^{+} = \frac{I_k}{I_p} - I^{+}$$

where I_p is the current of the primary beam on the free surface,
I^+ is the current in the transmitted-ion collector.

We studied emitters made of various thin films by passing Li^+, Na^+, and K^+ ions through them. The thin films were obtained by depositing the vapor on nickel grids with an 80 to 90% transparency on top of a previously-deposited film of nitrocellulose, which was subsequently removed. An emitting layer of MgO was prepared by the combustion of magnesium in air and deposition of the smoke on an aluminum foil 300 Å thick. With this method of obtaining the films, fairly loose layers of magnesium oxide were obtained, and their thickness was estimated by weighing.

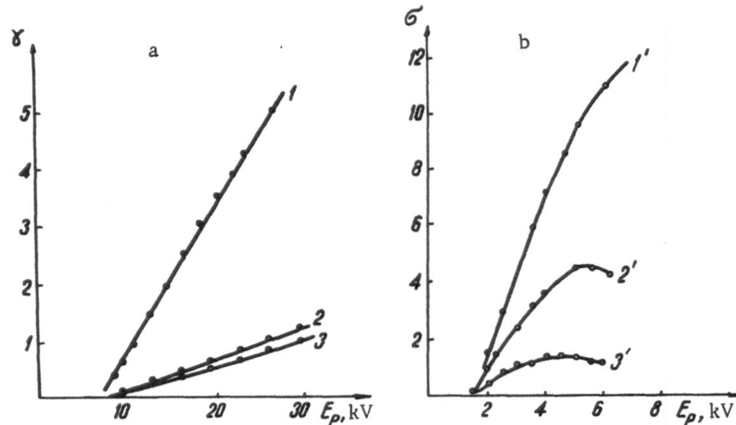

Fig. 2. Secondary-emission coefficients as functions of the accelerating voltage: a) $\gamma(E_p)$ due to Li^+ ions; b) $\sigma(E_p)$ due to electrons; 1) and 1′) for MgO (90 kg/cm²) on an Al substrate (d = 300 Å); 2) and 2′) for KCl (d = 400 Å) on an Al substrate (d = 300 Å); 3) and 3′) for Al (300 Å thick).

Measurement of the Secondary-Emission Coefficients in Transmission

The experimental data presented in Fig. 2 are of greatest interest. We see from Fig. 2a that the $\gamma(E_p)$ relationship obtained on firing 10- to 30-keV Li^+ ions is of a linear nature. We observe from the course of the straight lines that the processes determining ion–electron emission must clearly take place in the surface layers of the target (in any case at depths considerably smaller than the thickness of the films under consideration).

It is well known that in such experiments the somewhat distorted results are due to the presence of various adsorbed gases and vapors on the film. In order to obtain reasonably reliable quantitative results these must therefore be removed first by heating the film to an appropriate temperature. Thus it was shown earlier [7] that in order to remove various gases and oil vapor settling on the surface of the films it was sufficient to heat the latter to a temperature of the order of 200° C. In the course of bombarding the films with ions and electrons, they are heated to temperatures of the order of several hundreds of degrees, sufficient to remove adsorbed gases and vapors from the film surface.

Figure 2 (curves 2 and 3) shows that there is a slight difference between the secondary-emission coefficients obtained from KCl and Al films, respectively, on bombarding with Li^+ ions; on bombarding the same emitters with electrons (curves 2′ and 3′) a considerable difference is apparent. Such results may be explained either by considering that the work function of the metal surface (in the case of the adsorption of KCl on Al) is not a decisive factor in the kinetic extraction of electrons for a fairly high degree of coverage ($\Theta \gg 1$), or else by the rise in the temperature of the KCl-carrying emitter when the energy of the incident ions increases from 10 to 30 keV.

It is well known that, for an alkali halide of the KCl type, the phonon losses increase with rising temperature; the mechanism of scattering at lattice vibrations may become dominant and result in a lower γ. The scattering of excited electrons at defects may also play an important part; there may be as many as 10^{21} defects/cm³ in KCl films.

Highly-efficient emitters based on porous layers of MgO were obtained by N. L. Yasnopol'skii et al. in a study of secondary emission amplified by an electric field [9].

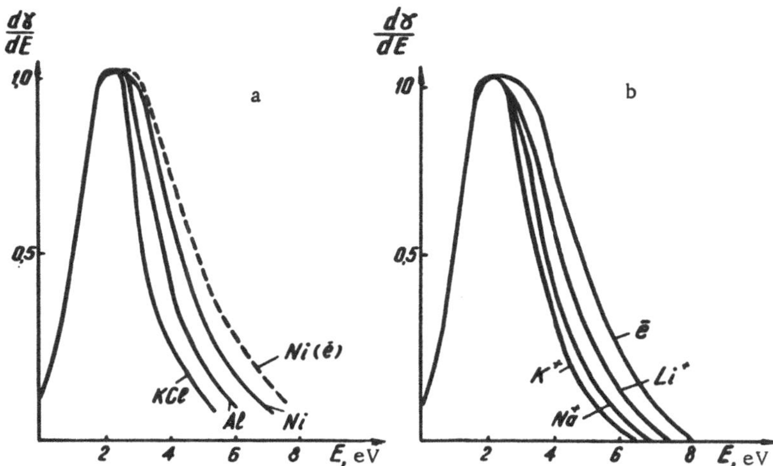

Fig. 3. Energy-distribution curves of secondary electrons: a) in relation to the substrate material on transmitting 20-keV Na^+ ions; b) in relation to the form of the incident particle. Emitter thicknesses as in Fig. 2.

We also made some experiments with emitters of this type and found (Fig. 2, curve 1) that γ reached a value of 5 for 25-keV Li^+ ions and a collector potential of 80 V. Increasing the collector potential to 250 V failed to produce self-sustained emission.

The results confirm that field-amplified secondary emission may be explained by the multiplication of "nascent" secondary electrons created by the primary beam in pores of the coating, while the potential on the collector aids their release into the surrounding vacuum. This explanation is in agreement with the emission mechanism suggested [9-12].

Energy Distribution of the Electrons Ejected by Ions and Electrons Passing through Thin Foils

It is found that varying the energy of the ions from 10 to 35 keV has no effect on the electron energy distribution (within the limits of experimental error).

Figure 3 shows the energy distributions (normalized to the same height) of electrons ejected from KCl, Al, and Ni films on passing 20-keV Li^+, Na^+, and K^+ ions and 44-keV electrons through them. We see from the curves of Fig. 3a that the width of the distribution curve depends on the film material. Films of potassium chloride give the sharpest and nickel films the widest distribution. In the case of potassium chloride the narrow distribution is sometimes accompanied by a wider one, apparently associated with the charging of the surface. In order to eliminate possible discrepancies in the energy distribution, only spectra obtained in an irradiation period of no longer than 30 sec should thus be considered.

Figure 3b shows the relation between the secondary-electron energy distribution obtained from an aluminum film (d = 300 Å) and the form of the bombarding particles. The number of secondary electrons with comparatively high energies diminishes on passing from lithium to potassium ions (i.e., the number of high-energy secondary electrons diminishes with increasing mass of the incident ion).

Our experimental results also show that the widest energy distributions are obtained by firing electrons through the films.

We also obtained a number of secondary-electron distribution curves from porous magnesium oxide films. The energy distribution of the electrons should be affected by the charging of the film surface. As a result of the escape of secondary electrons, the maximum positive charge density resides on the surface. As a result of this, the internal field intensifies the motion of secondary electrons in the direction of the escape surface. For a conducting substrate, the maximum surface potential equals the potential of the collector. In this case the electric field strength in the layer will be given by the expression E_1/d, where d is the thickness of the layer.

In order that the escape surface should become charged to the potential of the collector and a constant value of γ established, a certain finite time interval t is required; during this period a continuous rise in the secondary emission takes place. The value of t may be determined from the expression relating the charge to the voltage of a parallel condenser:

$$t = \frac{DE_1 \cdot 10^{-13}}{d(\bar{\gamma} - 1) I_p}$$

where $\bar{\gamma}$ is the mean secondary-emission coefficient during the time t (in sec), I_p is the beam current density (A/cm^2), d is the distance (cm), E_1 is the collector voltage (V), and D is the dielectric constant, which may be taken as unity, since 95% of the volume of the layer is occupied by vacuum, i.e., for current densities of 10^{-7} to 10^{-8} A/cm^2 the emitters are charged almost instantaneously. Owing to the considerable dependence of the secondary-electron energy distribution on the charge covering the emitting side of the film, the accuracy of such measurements was not particularly high.

References

1. A. Becker, Ann. Phys., II, 357 (1927).
2. H. Bush, Ann. Univ. Saare, 4 (1953).
3. V. G. Butkevich and M. M. Butslov, Radiotekhnika i Élektronika, 3:355 (1958).
4. E. J. Sternglass, Rev. Sci. Inst., 26:1202 (1955).
5. W. Dietrich and H. Seller, Z. Phys., 157:576 (1960).
6. U. A. Arifov, A. Kh. Ayukhanov, and S. V. Starodubtsev, Zh. Éksp. Teor. Fiz., 26:714 (1954).
7. A. E. Ennos, Brit. J. Appl. Phys., 5:27 (1954).
8. B. Petzel, Ann. Phys., 6:55 (1960); 9:44 (1961).
9. N. L. Yasnopol'skii, N. A. Karelina, and V. S. Malysheva, Radiotekhnika i Élektronika, 5(10):1747 (1960).
10. N. L. Yasnopol'skii, N. A. Karelina, and V. S. Malysheva, Radiotekhnika i Élektronika, 6(1):146 (1961).
11. N. L. Yasnopol'skii, N. A. Karelina, and V. S. Malysheva, Radiotekhnika i Élektronika, Vol. 7, No. 7 (1962).
12. N. L. Yasnopol'skii, N. A. Karelina, and V. S. Malysheva, Radiotekhnika i Élektronika, 7(9):1657 (1962).

INFLUENCE OF ION BOMBARDMENT ON THE PHYSICAL AND PHYSICOCHEMICAL PROPERTIES OF SILICON

Sh. A. Ablyaev, V. P. Chirva, and V. V. Arsenin

The method of ionic doping is now finding wider and wider applications in various fields of semiconductor electronics. Although the method has been intensively studied, very little published data exist.

In recent years several experimental investigations into the range of ions and the formation of p–n junctions in silicon irradiated with Cs, Li, B, and several other types of ions at energies up to 10 keV have appeared [1-6]. In this paper we shall present some preliminary results of an investigation into the changes taking place in certain physical and physicochemical properties of p- and n-type silicon irradiated with Cs, Li, B, Sb, and Si ions at energies up to 50 keV.

In order to obtain the positive ions we used an ion source based on the surface ionization of atoms and a universal ion source developed in our own laboratory.

The preparation of the samples reduced to the chemical polishing of the silicon surface in a standard polishing etchant. On the other (ground) side, an ohmic nickel contact was deposited chemically. Using an ultrasonic machine, discs 3 mm in diameter and 0.6 mm thick were cut. The samples were degreased, washed in deionized water and dried; then they were subjected to ion bombardment in a vacuum of the order of 10^{-6} mm Hg at room temperature, using a variety of doses and ions of various energies. The irradiation of the silicon surface was carried out parallel to the (111) crystallographic direction. In order to ensure better cleaning of the silicon surface, a low-voltage glow discharge was ignited over the latter for several minutes before irradiating in the ion-beam tube.

The layer resistance was measured by a four-probe method, the volt–ampere characteristics in the usual manner (a layer of cement made from a mixture of In, Ga, and Zn was introduced onto the polished surface of the irradiated silicon by light rubbing). In order to avoid the heating of the samples during irradiation, the density of the ion beams employed never exceeded 2 to 3 $\mu A/cm^2$.

As a result of the bombardment of the silicon with positive ions of medium energies, a thin amorphous layer was formed on the surface; the electrical resistance of this was considerably higher than that of the original material. The state of the irradiated surface was monitored, not only by the probe, but also by an optical method, using a Linnik interference microscope. The experiments showed that the irradiated places on the silicon surface acquired a

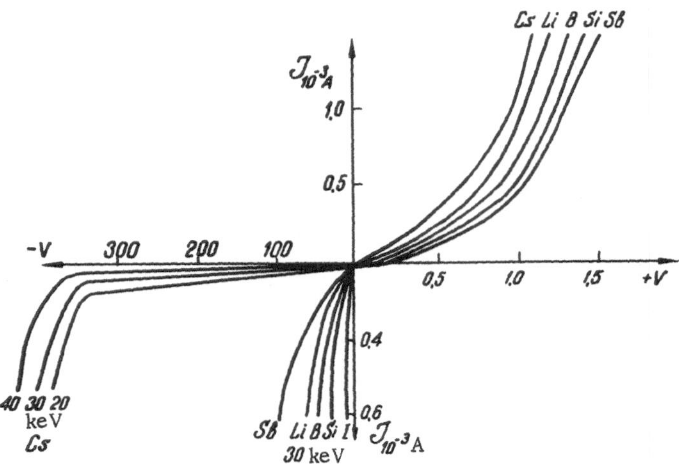

Fig. 1. Volt–ampere characteristics for diodes obtained by ion bombardment: 1) reverse branch of the V–A characteristic for diodes annealed at 500°C in a hydrogen atmosphere for 0.5 h; ρ = 0.5 Ω·cm for Li, B, Sb, and Si ions and ρ = 100 Ω·cm for Cs ions.

distinct color, the intensity of which varied with the dose of radiation. This was evidently associated with the formation of a thin surface layer possessing different physicochemical properties from those of the original material. The thickness of the layer could be determined from the coloring (0.06 to 0.08 μ). Various organic solvents (benzene, dichloroethane, ethyl alcohol, etc.) and acids had no effect on the surface layer formed after irradiation. However, there was an increase in the solubility of the irradiated parts in hydrofluoric acid and in the standard polishing etchant CP-4.

The bombardment of a silicon surface with ions of various elements leads to the formation of an inversion layer [7], the depth of this being determined by the energy of the ions employed. The volt–ampere characteristics of the resultant p–n junctions are presented in Fig. 1. The best values of rectification factor were obtained on bombarding the silicon surface with Cs and Sn ions. The thermal annealing of the irradiated samples in a hydrogen atmosphere at 500°C for 0.5 h leads to a sharp reduction in the rectification factor.

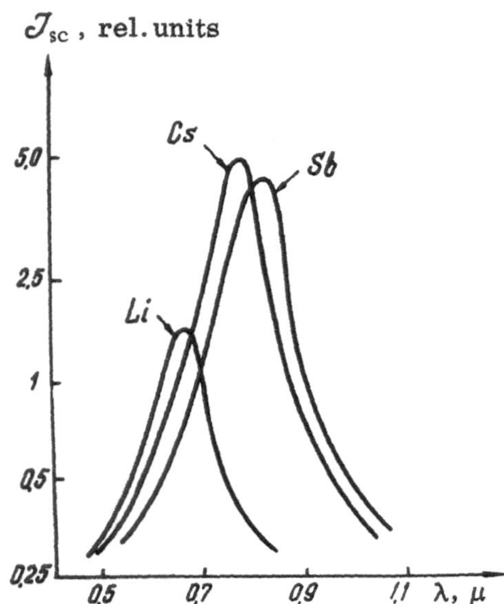

Fig. 2. Spectral characteristics of photodiodes in the short-circuit mode: E = 30 keV, D = 100 μC/cm^2.

In considering the influence of ion bombardment on the electrophysical properties of silicon, it is important to allow for the development of two kinds of defect: essentially radiation defects (individual displaced atoms and larger aggregates of these), and interstitial defects, constituting ions of foreign material penetrating into the lattice and acting as ordinary chemical impurities. Both these types of defects cause a change in carrier concentration and also lead to a change in carrier mobility as a result of the creation of new scattering centers. The observed changes in the electrical properties after irradiation depend on the relation between the concentration and efficiency of the energy levels introduced during the irradiation and those existing prior to the latter. The effect of the type of ion on the changes in the electrical properties can only be detected under conditions in which the interstitial defects play the main part in the observed effects. Such conditions may be created by carrying out the irradiation at high temperatures, at which complete annealing of the radiation defects occurs, or by thermal annealing after the ion bombardment of the silicon. In studying the electrical properties of the surface layer of silicon subjected to ion irradiation, great difficulties arise from the thinness of the surface layer and the complexity of the procedure for separating surface and volume effects.

Samples with stable p–n junctions were used for the manufacture of some of our photocells [8, 9]. The best of these were photodiodes obtained by bombarding the silicon surface with Cs and Sb ions. The spectral characteristics of photocells annealed in hydrogen at 500°C for 0.5 h (in the short-circuit mode) are presented in Fig. 2.

The photosensitivity of the photoreceivers in question is several tens of times higher than that of diodes not thermally annealed. With increasing energy of the incident particles (in the case of cesium ions), the position of the maximum of the spectral sensitivity moves slightly in the long-wave direction. After annealing in hydrogen, the position of the spectral-sensitivity maximum obtained for 40 keV ions lies at 0.87 μ. Further increasing the period of thermal annealing of the irradiated samples leads to a gradual fall in the sensitivity of the photodiodes and to the partial vanishing of the color from the irradiated surface.

References

1. M. M. Bredov, Dokl. Akad. Nauk SSSR, 113(4):795 (1957).
2. M. M. Bredov, Fiz. Tverd. Tela, No. 4, 2, 562-564 (1962).
3. J. O. McCaldin, J. Phys. Chem. Solids, 24(9):1073 (1963).
4. D. B. Medved et al., Appl. Phys. Letters, 3(12):213 (1963).
5. P. V. Pavlov et al., Fiz. Tverd. Tela, Vol. 81, No. 3 (1966).
6. M. Waldner and P. E. McQuaid, Solid-State Electronics, Pergamon Press, No. 7, pp. 925-931 (1964).
7. Sh. A. Ablyaev, V. M. Mikhaélyan, and V. P. Chirva, Izv. Akad. Nauk Uzbek. SSR, Ser. Fiz., No. 2, p. 93 (1968).
8. Sh. A. Ablyaev and V. P. Chirva, Geliotekhnika, No. 6, p. 9 (1967).
9. Sh. A. Ablyaev and V. P. Chirva, Geliotekhnika, No. 2, p. 45 (1968).

COMPUTER STUDY OF THE PROTON–ELECTRON–POSITRON SYSTEM BY A VARIATIONAL METHOD

P. U. Arifov and V. M. Mal'yan

The variational method [1] of studying many-particle systems lies in minimizing the energy functional

$$E = \frac{\int \psi^* \hat{H} \psi d\tau}{\int |\psi|^2 d\tau} = \min \tag{1}$$

by means of an appropriately-constructed test (trial) function (based on physical and other considerations), constituting a certain approximation to the true eigenfunction of the ground state of the system; the resultant minimum energy E_{min} gives the upper limit to the eigenvalue of the energy of this state.

In order to determine the function best suited to the physical essence of the system under consideration, we examined test functions of various types on an electronic computer, namely:

$$\left.\begin{aligned} \psi_1 &= e^{-\lambda_1 r_1 - \lambda_2 r_2} \\ \psi_2 &= e^{-\alpha S - \beta t}(1 + cu) \\ \psi_3 &= r_2^{\sigma} e^{-\alpha r_1 - \gamma r_{12}} \\ \psi_4 &= e^{-\alpha S - \beta t - \gamma u} \\ \psi_5 &= e^{-\alpha r_1 - \gamma r_{12}} + \beta e^{-\gamma r_1 - \alpha r_{12}} \end{aligned}\right\} \tag{2}$$

where λ_1, λ_2, α, β, c, γ, σ constitute variational parameters,
r_1, r_2 are the distances from the electron and positron to the nucleus;
r_{12} is the distance from the positron to the electron,
s, t, u are the Hylleraas variables [1] $s = r_1 + r_2$; $t = r_1 - r_2$; $u = r_{12}$.

In the variables r_1, r_2, r_{12} the energy functional to be minimized takes the following form (in atomic units)

$$E = N^{-1} \left\{ \int d\tau \left[\frac{1}{2} \left(\frac{\partial \psi}{\partial r_1} \right)^2 + \frac{1}{2} \left(\frac{\partial \psi}{\partial r_2} \right)^2 + \left(\frac{\partial \psi}{\partial r_{12}} \right)^2 + \right.\right.$$

$$\left.\left. + \frac{1}{2} \frac{r_{12}^2 + r_1^2 - r_2^2}{r_1 r_{12}} \frac{\partial \psi}{\partial r_1} \frac{\partial \psi}{\partial r_{12}} + \frac{1}{2} \frac{r_{12}^2 + r_2^2 - r_1^2}{r_2 r_{12}} \frac{\partial \psi}{\partial r_2} \frac{\partial \psi}{\partial r_{12}} \right] + \int d\tau \psi^2 \left(-\frac{1}{r_1} - \frac{1}{r_{12}} + \frac{1}{r_2} \right) \right\} \tag{3}$$

where

$$N = \int \psi^2 d\tau$$

and in Hylleraas variables

$$E = N^{-1}\int_0^{\infty} ds \int_0^{S} du \int_{-u}^{u} dt \left\{ u(s^2 - t^2)\left[\left(\frac{\partial\psi}{\partial s}\right)^2 + \left(\frac{\partial\psi}{\partial t}\right)^2 + \left(\frac{\partial\psi}{\partial u}\right)^2\right] + \right.$$

$$\left. + 2\frac{\partial\psi}{\partial u}\left[(u^2 - t^2)\, s\frac{\partial\psi}{\partial s} + t(s^2 - u^2)\frac{\partial\psi}{\partial t}\right] + (4ztu - s^2 + t^2)\psi^2 \right\} \tag{4}$$

Here

$$N = \int_0^{\infty} ds \int_0^{S} du \int_{-u}^{u} dt\, \psi^2 u(s^2 - t^2)$$

Investigations based on trial functions ψ_1 (describing the independent motion of the light particles) and ψ_2 and ψ_3 (allowing for the polarization interaction between antiparticles and positron coupling respectively) yielded no normal minimum ("pit" or "well") [2] for acceptable values of the variational parameters. These trial functions, constructed on the assumption that the pe^+e^- system was analogous to the helium atom, were characterized by fields of energy values E possessing one (ψ_2) or two (ψ_3) descents to minima, reached at the very limit of the acceptable values of the variational parameters, but not characterizing a localized pe^+e^- system [2].

It follows from the calculations thus carried out that the pe^+e^- system should more correctly be regarded as an analog of the negative hydrogen ion, rather than the helium atom. In order to secure an indirect verification of this proposition, we studied the function ψ_4 (Fig. 1a), it being well known [1] that the simplest functions of the type ψ_1, ψ_2, ψ_3 fail to describe the stable bound state of the negative hydrogen ion.

For equal values of the parameters α and β, two regions of descent to a minimum appear, as in the case of the earlier function ψ_3. These are reached at the boundary of the acceptable range of parameter values describing states of the pe^+e^- system such that one of the positive particles is not localized close to the others. For unequal values of the parameters α and β, ($\alpha \neq \beta$), as in the case of the function ψ_2, single region of descent to a minimum is obtained (Fig. 1b). With increasing γ the descent becomes flatter and the minimum value reached at the boundary increases.

We also considered a function ψ_5 symmetrized with respect to the coordinates of the particles with the same charge sign, in the same way as that employed earlier when considering the ground state of the negative hydrogen ion [3]. The parameter β allowed for the fact that the proton and positron were of different natures.

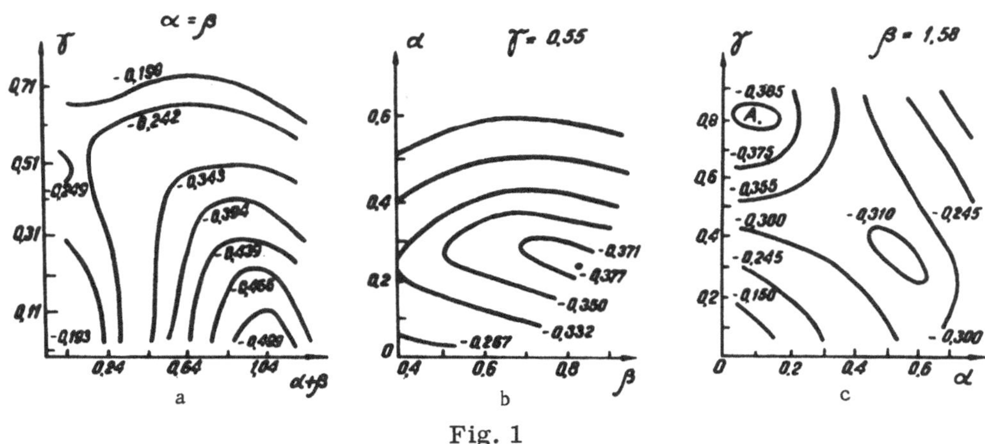

Fig. 1

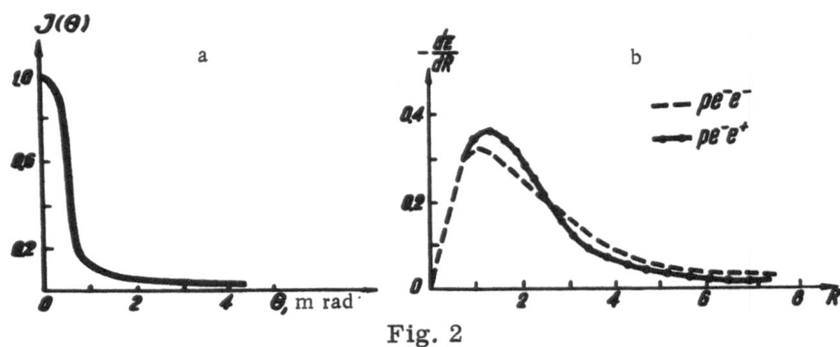

Fig. 2

We see from the calculations that, for parameter values of $\beta < 0.5$ and $\beta > 1.6$, as in the case of the functions ψ_3 and ψ_4, there are two regions of descent, while for values of $0.5 < \beta < 1.6$ there are two clearly-expressed minima ("wells"), which are arranged symmetrically for $\beta = 1$ and are characterized by the same energy value of $E_{min} = -0.314$ at. units. The values of the parameters in the region of the minima characterize a completely localized pe^+e^- system. The minimum energy value is found at the point A (Fig. 1c) with the coordinates $\alpha = 0.12$, $\gamma = 0.78$, $\beta = 1.58$; $E_{min} = -0.38843$ at. units.

The resultant value of our upper estimate for the eigenvalue of the energy corresponding to the ground state of the pe^+e^- system is much lower than that obtained by Shmelev [4] ($E_{min} = -0.2815$ at. units) and Din-Van-Hoang [5] ($E_{min} = -0.2838$ at. units). Our value indicates the stability of the system relative to decomposition into a proton and positronium and also into a positron and a hydrogen atom in any excited state, although the system is unstable relative to decomposition into a positron and a hydrogen atom in the ground state.

The lifetime of the positron in the state described by the symmetrized function ψ_5 was similar to the annihilation time in a free atom of positronium ($\tau_{pe^- e^+} = 1.16975 \cdot 10^{-10}$ sec). The angular correlation function of the annihilation quanta is distinguished by a narrow distribution, some 95% of the decompositions being included within a range of 3 mrad (Fig. 2a).

On considering the results of the calculation and comparing the electron-density distributions of pe^-e^- and pe^-e^+, we see (Fig. 2b) that the replacement of an electron by a positron produces a considerable deformation in the distribution of negative charge.

Let us consider the identification of the state described by the function ψ_5 in the annihilation experiment. Although this state is not stable with respect to the ejection of a positron, in order to secure a proper interpretation of experimental data we must know the degree of instability, i.e., the time required for pe^+e^- to decay into e^+ and H. If the system decomposes into its components in a time two or three orders of magnitude shorter than the annihilation time (10^{-10} sec), then it is impossible to observe the existence of the state described by the function ψ_5 experimentally. If, however, the lifetime of the pe^+e^- relative to decay into its corresponding components is greater than, of the same order as, or under certain conditions even one or two orders smaller than the annihilation time, then the annihilation characteristics of the bound pe^+e^- system should appear in the annihilation experiments. This state may be identified, for example, from the form of the angular-correlation curve, pe^+e^- being characterized by a narrow distribution. The foregoing proposition is based on the fact that, in a gas medium, the process of retardation by elastic collisions is fairly prolonged, i.e., in the time required for the positron to be slowed down to thermal velocities, the pe^+e^- system may develop and decay repeatedly, while at a certain moment (if there are enough of such events) the pe^+e^- will decay by way of an annihilation process. In this connection it will in future investigations be interesting to study the questions of positron retardation considered earlier [6] in more detail for specific gas media.

Appendix

I. In studying the function ψ_2 we minimized

$$E=\frac{\eta\xi}{4}+\frac{b^2}{2}+2N^{-1}b\Big\{\big[b\xi^2\eta^5(4b^2+6b\xi+6\xi^2)+b\xi^5\eta^2(-4b^2+6b\eta-$$

$$-6\eta^2)\big](2c-c^2+zb)-\Big[2acb^2\xi^2\eta^5\Big(-b-\xi+\frac{\xi^3}{a^2}\Big)+2acb^2\xi^5\eta^2\Big(b-\eta+\frac{\eta^3}{a^2}\Big)\Big]-$$

$$-\big[3ac^2b\xi\eta^5(-2b^2-2b\xi-\xi^2)+3ac^2b\xi^5\eta(2b^2-2b\eta+\eta^2)\big]+\Big[2cb^2\eta^5\xi^2<\frac{\xi^3}{a}\Big(\frac{b^2}{a^2}-2\Big)+b^2+2b\xi+2\xi^2>+$$

$$+2cb^2\eta^2\xi^5<\frac{\eta^3}{a}\Big(\frac{b^2}{a^2}-2\Big)+b^2-2b\eta+2\eta^2>\Big]+\big[3c^2b\xi\eta^5(2b^3+4b^2\xi+5b\xi^2+5\xi^3)+$$

$$+3c^2b\xi^5\eta(2b^3-4b^2\eta+5b\eta^2-5\eta^3)\big]+\Big[8zcb\xi\eta^5\Big(3b^3+4b^2\xi+4b\xi^2+4\xi^3-4\frac{\xi^4}{a}\Big)+$$

$$+8zcb\xi^5\eta\Big(-3b^3+4b^2\eta-4b\eta^2+4\eta^3-\frac{4\eta^4}{a}\Big)+\big[12zc^2\eta^5(4b^4+5b^3\xi+5b^2\xi^2+5b\xi^3+5\xi^4)+$$

$$+12zc^2\xi^5(-4b^4+5b^3\eta-5b^2\eta^2+5b\eta^3-5\eta^4)\big]+\Big[b^2\xi^3\eta^5<2b+3\xi+\frac{\xi^2}{a}\Big(\frac{b^2}{a}-3\Big)>+$$

$$+b^2\xi^5\eta^3<-2b+3\eta+\frac{\eta^2}{a}\Big(\frac{b^2}{a^2}-3\Big)>\Big]+\Big[2c<c\xi\eta^5\Big(6b^3+9b^2\xi+10b\xi^2+10\xi^3+\frac{\xi^4}{a}\Big(\frac{b^2}{a^2}-10\Big)\Big)+$$

$$+c\xi^5\eta\Big(-6b^3+9b^2\eta-10b\eta^2+10\eta^3+\frac{\eta^4}{a}\Big(\frac{b^2}{a^2}-10\Big)\Big]\Big]\Big\}$$

where

$$N=-\big[2b<b\xi^2\eta^5(4b^2+6b\xi+6\xi^2)+b\xi^5\eta^2(-4b^2+4b\eta-6\eta^2)\big]-\Big[8b<c\xi\eta^5(6b^3+9b^2\xi+10b\xi^2+$$

$$+10\xi^3+\frac{\xi^4}{a}\Big(\frac{b^2}{a^2}-10\Big)\Big)+c\xi^5\eta\Big(-6b^3+9b^2\eta-10b\eta^2+10\eta^3+\frac{\eta^4}{a}\Big(\frac{b^2}{a^2}-10\Big)\Big)>\Big]+$$

$$+\big[c^2\eta^5(-96b^4-144b^3\xi-168b^2\xi^2-180b\xi^3-180\xi^4)+c^2\xi^5(96b^4-144b^3\eta+168b^2\eta^2-180b\eta^3+180\eta^4)\big]$$

$$a=2\alpha;\quad b=2\beta;\quad \xi=a+b;\quad \eta=a-b$$

II. In studying ψ_3 we minimized

$$E(\alpha,\gamma,\sigma)=\frac{1}{N}\sum_{n=1}^{10}I_n;\qquad N=I_{11}$$

$$I_n=R_n(\alpha,\gamma,\sigma)\,d_n!\sum_{k=0}^{d_n}I_k$$

$$I_k=\frac{1}{\gamma_n^{k+1}}\sum_{i=0}^{d_{n-k}}I_i$$

$$J_i=\frac{(l_n+i)!}{i!(d_n-k-i)!}\left\{\frac{(c_n+d_n-k-i)!}{(\alpha_n+\gamma_n)^{c_n+d_n-k-i+1}}\left(\frac{(-1)^l}{(-\gamma_n)^{l_n+l+1}}-\frac{1}{(\gamma_n)^{l_n+l+1}}\right)+\sum_{j=0}^{l_n+l}\varphi_j\right\}$$

$$\varphi_j=\frac{(c_n+d_n-k+l_n-j)!}{(l_n+i-j)!\,\alpha_n^{c_n+d_n-k+l_n-j+1}}\left(\frac{(-1)^{d_n-k-i}}{(\gamma_n)^{j+1}}-\frac{(-1)^l}{(-\gamma_n)^{j+1}}\right)$$

$$R_1=R_{11}=1;\ R_2=\frac{\alpha^2}{2}+\gamma^2;\ R_3=-\frac{\sigma\gamma}{2}-1;\ R_4=R_6=\frac{\alpha\gamma}{2}$$

$$R_5=-1;\ R_7=\frac{c\gamma}{2};\ R_8=-\frac{\sigma\gamma}{2};\ R_9=\frac{\sigma^2}{2};\ R_{10}=-\frac{\alpha\gamma}{2}$$

$$c_1 = c_2 = c_2 = c_8 = c_9 = c_{11} = 1; \quad c_4 = c_5 = c_{10} = 0; \quad c_6 = 2; \quad c_7 = 3.$$

$$l_1 = 2\sigma; \quad l_2 = l_3 = l_5 = l_6 = l_{11} = 2\sigma + 1; \quad l_7 = l_8 = l_9 = 2\sigma - 1; \quad l_{10} = 2\sigma + 3$$

$$d_1 = d_2 = d_5 = d_9 = d_{11} = 1; \quad d_3 = d_6 = d_7 = d_{10} = 0; \quad d_4 = d_8 = 2$$

$$\alpha_1 = \alpha_2 = \cdots \alpha_{11} = 2\alpha; \quad \gamma_1 = \gamma_2 = \cdots = \gamma_{11} = 2\gamma$$

III. In studying the ψ_4 we minimized

$$E(\alpha, \beta, \gamma) = \left(\sum_{k=1}^{9} l_k I_k\right) \cdot N^{-1}; \qquad N = I_{10} + I_{11}$$

$$I_k = \sum_{j=0}^{b_k} \frac{b_k!}{(b_k - j)!\,\beta^{j+1}} \sum_{m=0}^{c_k + b_k - j} \frac{(c_k + b_k - j)!}{(c_k + b_k - j - m)!\,(\gamma + \beta)^{m+1}} \frac{(a_k + c_k + b_k - j - m)!}{(\gamma + \beta)^{a_k + c_k + b_k - j - m + 1}} - \sum_{j=0}^{b_k} \frac{b_k!}{(b_k - j)!\,\beta^{j+1}} \times$$

$$\times \frac{(c_k + b_k - j)!}{(\alpha + \beta)^{c_k + b_k - j + 1}} \cdot \frac{a_k!}{\alpha^{a_k + 1}} - \sum_{j=0}^{b_k} (-1)^{b_k - j} \left| \frac{b_k!}{(b_k - j)!\,\beta^{j+1}} \sum_{m=0}^{c_k + b_k - j} \frac{(c_k + b_k - j)!}{(c_k + b_k - j - m)!\,(\gamma - \beta)^{m+1}} \times \right.$$

$$\times \frac{(a_k + c_k + b_k - j - m)!}{(\alpha + \beta + \gamma)^{a_k + c_k + b_k - j - m + 1}} + \sum_{j=0}^{b_k} (-1)^{b_k - j} \frac{b_k!}{(b_k - j)!\,\beta^{j+1}} \frac{(c_k + b_k - j)!}{(\gamma - \beta)^{c_k + b_k - j + 1}} \cdot \frac{a_k!}{\alpha^{a_k + 1}}$$

$$l_1 = \alpha^2 + \beta^2 + \gamma^2; \quad l_2 = -l_1; \quad l_3 = 2\alpha\gamma; \quad l_4 = -2\alpha\gamma; \quad l_5 = 2\beta\gamma$$

$$l_6 = -2\beta\gamma; \quad l_7 = 4z; \quad l_8 = l_{11} = -1; \quad l_9 = l_{10} = 1$$

$$a_1 = a_5 = a_8 = a_{10} = 2; \quad a_2 = a_6 = a_7 = a_9 = a_{11} = 0; \quad a_3 = a_4 = 1$$

$$b_1 = b_3 = b_8 = b_{10} = 0; \quad b_5 = b_6 = b_7 = 1; \quad b_2 = b_4 = b_9 = b_{11} = 2$$

$$c_1 = c_2 = c_7 = c_{10} = c_{11} = 1; \quad c_3 = c_6 = 2; \quad c_4 = c_5 = c_8 = c_9 = 0$$

IV. The electron density was calculated from the formula

$$-\frac{dz}{dr} = \frac{1}{N} \sum_{n=1}^{3} I_n(r)$$

$$I_n(r) = d_n!\, e^{-\alpha_n r} \sum_{k=0}^{d_n} \frac{1}{(\gamma_n)^{k+1}} \sum_{i=0}^{d_n - k} \frac{(l_n + i)!}{(d_n - k - i)!\,i!} \sum_{j=0}^{l_n + i} \frac{1}{(l_n + i - j)!} r_1^{c_n + d_n - k + l_n - j} \times$$

$$\times \left(\frac{(-1)^{d_n - k - i}}{(\gamma_n)^{j+1}} - \frac{(-1)^i}{(-\gamma_n)^{j+1}} \right) R_n(\beta)$$

$$c_1 = c_2 = c_3 = 1; \quad d_1 = d_2 = d_3 = 1; \quad l_1 = l_2 = l_3 = 1$$

$$\alpha_1 = 2\alpha; \quad \alpha_2 = 2\gamma; \quad \alpha_3 = (\alpha + \gamma)$$

$$\gamma_1 = 2\gamma; \quad \gamma_2 = 2\alpha; \quad \gamma_3 = (\alpha + \gamma)$$

$$R_1(\beta) = 1; \quad R_2(\beta) = \beta^2; \quad R_3(\beta) = 2\beta$$

All the calculations carried out in the present investigation were based on standard programs for calculating expressions of the form:

$$I = \int_0^\infty dr_1 \int_0^{r_1} dr_2 \int_{r_1 - r_2}^{r_1 + r_2} dr_{12}\, e^{-\alpha r_1 - \beta r_2 - \gamma r_{12}} \cdot r_1^c r_2^l r_{12}^d + \int_0^\infty dr_1 \int_{r_1}^\infty dr_2 \int_{r_2 - r_1}^{r_2 + r_1} dr_{12}\, e^{-\alpha r_1 - \beta r_2 - \gamma r_{12}} r_1^c r_2^l r_{12}^d =$$

$$= d! \sum_{k=0}^{d} \frac{1}{\gamma^{k+1}} \sum_{i=0}^{d-k} \frac{(l+i)!}{(d-k-i)!\,i} \left\{ \frac{(c + d - k - i)!}{(\alpha + \gamma)^{c + d - k - i + 1}} \left[-\frac{1}{(\beta + \gamma)^{l+i+1}} + \frac{(-1)^i}{(\beta - \gamma)^{l+i+1}} \right] + \right.$$

$$\left. + \sum_{j=0}^{l+1} \frac{(c + d - k + l - j)!}{(l + i - j)!\,(\alpha + \beta)^{c + d - k + l - j + 1}} \left[\frac{(-1)^{d-k-i}}{(\beta + \gamma)^{j+1}} - \frac{(-1)^i}{(\beta - \gamma)^{j+1}} \right] \right\}$$

$$I=\int_0^\infty ds\int_0^s du\int_{-u}^{u} dt s^a t^b u^c e^{-\alpha s-\beta t-\gamma u}=\sum_{j=0}^{b}\frac{b!\,(c+b-j)!}{\beta^{j+1}(b-j)!}\left\{\sum_{m=0}^{c+b-j}\frac{(a+b+c-j-m)!}{(b+c-j-m)!}\times\right.$$

$$\times\left[\frac{(1)}{(\gamma+\beta)^{m+1}(\gamma+\beta+\alpha)^{a+c+b-j-m+1}}-\frac{(-1)^{b-j}}{(\gamma-\beta)^{m+1}(\gamma-\beta+\alpha)^{a+c+b-j-m+1}}\right]-$$

$$\left.-\frac{a!}{\alpha^{a+1}}\left[\frac{1}{(\gamma+\beta)^{b+c-j+1}}-\frac{(-1)^{b-j}}{(\gamma-\beta)^{b+c-j+1}}\right]\right\}$$

programed in the M-20, Minsk-14, and Minsk-22 electronic computer codes.

References

1. H. Bethe and E. Solpiter, Quantum Mechanics of Atoms with One and Two Electrons [Russian translation], Fizmatgiz, Moscow (1960).
2. P. U. Arifov, Dokl. Akad. Nauk Uzbek. SSR, 3:18 (1967).
3. T. Chandrasekhar, Astrophys. J., 100:176 (1944).
4. V. P. Shmelev, Zh. Éksp. Teor. Fiz., 37:458 (1959); 38:1528 (1960); Author's abstract of Candidate's Dissertation, Moscow State University (1960).
5. Din-Van-Khoang, Zh. Éksp. Teor. Fiz., 49:630 (1965); Author's abstract of Candidate's Dissertation, Belorussian State University (1965).
6. P. U. Arifov, V. I. Gol'danskii, and Yu. S. Sayasov, Izv. Akad. Nauk Uzbek SSR, Ser. Fiz.-Mat. Nauk, No. 5, p. 48 (1966).

ANNIHILATION OF POSITRONS AND THE ELECTRON SPECTRUM OF METALS

P. U. Arifov

It is well known [1] that in metals positrons are able to become fully thermalized before being annihilated. This means that the momentum spectrum of the electron–positron pairs restored from the angular correlation function of the annihilation gamma quanta in fact constitutes the electron-momentum spectrum, to a certain extent deformed by the presence of the positron which initiates the annihilation process in the metal.

Under conditions of point–line geometry, the solution to the problem of restoring the momentum spectrum of annihilating electron–positron pairs by reference to experimental data relating to the angular correlation of the annihilation quanta reduces to an integral equation of the Volterre type and of the first kind:

$$I(\alpha) = A\left\{ \int_{\alpha^2}^{\infty} \rho(w)\,dw - \frac{2^{3/2}}{\pi\Delta}\int_{\alpha^2}^{\infty} \rho(w)\sqrt{w-\alpha^2}\,dw - \frac{2^{3/2}}{\pi\Delta}\int_{\alpha^2+\frac{\Delta^2}{2}}^{\infty} \rho(w)\sqrt{w-\alpha^2-\frac{\Delta^2}{2}}\,dw + \right.$$

$$\left. + \frac{2}{\pi\Delta}\int_{\alpha^2+\frac{\Delta^2}{8}}^{\infty} \rho(w)\sqrt{w-\alpha^2-\Delta^2/8}\,dw - \frac{2^{3/2}}{\pi}\int_{\alpha^2+\frac{\Delta^2}{2}}^{\infty} \rho(w)\arctan\sqrt{\frac{w-\alpha^2-\Delta^2/2}{\Delta^2/2}}\,dw \right\} \tag{1}$$

where $\rho(W)$ is the unknown momentum spectrum, $w = p^2$;
p is the momentum of the pair,
$I(\alpha)$ is the experimental angular-correlation function,
A is a scale factor,
Δ is a constant characterizing the experimental apparatus [2].

The solution of Eq. (1) may be constructed in general form by using the method of successive approximations [2, 3]; however, it is better to take advantage of the potentialities of high-speed computers. In conformity with the algorithm proposed earlier [3], we developed a program for analyzing experimental angular-correlation data on the Minsk-14 computer. The results of an analysis of the experimental data of Stewart [4] are presented in Fig. 1. The points (including an indication of the error) constitute the results of Stewart, the broken curve is the first approximation of ρ_0, the dotted and dashed curve represents $-\rho = \rho_0 + \rho_1 + \rho_2$. The function ρ_0 was calculated by the Lanczos five-point method [5]

$$\rho_0 = -\frac{1}{A}\frac{dI}{d\alpha^2} \tag{2}$$

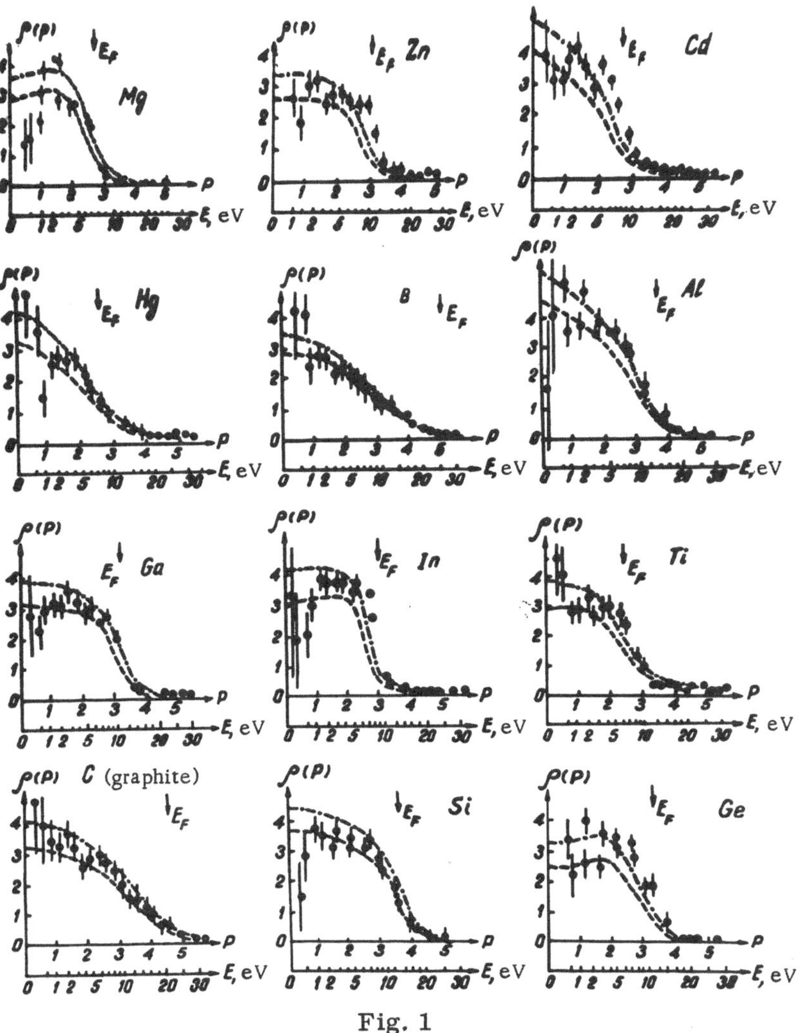

Fig. 1

based on approximating each group of five neighboring experimental points by a second-order parabola, obtained by the method of least squares. The next approximations [2, 3]

$$\rho_{n+1} = \int_{\alpha^2}^{\infty} K(w, \alpha^2)\, \rho_n(w)\, dw \tag{3}$$

are found by the trapezium method. Since the integrals being calculated contained a singularity, while the trapezium method failed to provide a strict estimation of the error in the calculations, the accuracy of the integral calculations was verified by using a stepped approximation of the functions ρ_i. The lower estimate agreed with the data resulting from the trapezium method, to within the plotting accuracy of the curves in the figure.

The computer calculations showed that the solutions found by the method of least squares converged very rapidly indeed for all the curves analyzed; the third and subsequent approximations altogether made a contribution of under 1%. Thus the program here proposed for analyzing the experimental angular-correlation data on the computer does in fact enable us to restore the electron–positron annihilation-pair spectra to a reasonable accuracy. The use of integral geometry for the experiment [6], together with machine analysis of the results, in our opinion offers the possibility of bringing the accuracy with which the momentum spectrum of the electrons in metals may be restored from annihilation experiments to the same level as that achieved when studying the electron spectrum of metals by analyzing soft x-ray emission [7]. We accordingly envisage the possibility of conducting a direct investigation into the perturbing effect of a positron on the electron states in a metal.

References

1. G. S. Zhdanov (ed.), Annihilation of Positrons in Solids [Russian translation] (Collection), IL, Moscow (1960).
2. P. U. Arifov, V. I. Gol'danskii, and Yu. S. Sayasov, Fiz. Tverd. Tela, 6:3118 (1964).
3. P. U. Arifov, Dokl. Akad. Nauk Uzbek SSR, 3:18 (1967).
4. A. Stewart, Canad. J. Phys., 35:168 (1957).
5. K. Lanczos, Practical Methods of Applied Analysis [Russian translation], Fizmatgiz, Moscow (1961).
6. P. U. Arifov, Dokl. Akad. Nauk Uzbek. SSR, 12:14 (1966).
7. F. Seitz, Modern Theory of Solids [Russian translation], GITTL, Moscow (1949).

SECONDARY ELECTRON EMISSION IN VARIOUS STATES OF AGGREGATION

U. A. Arifov and A. Kh. Kasymov

The secondary electron-emission coefficient of metals in various states of aggregation has already been studied [1-5]. However, the results of such investigations for the same material in different states of aggregation are often contradictory. Our present purpose is to elucidate the reasons for the discrepancies between existing experimental results and to provide a valid interpretation for the physical nature of this complicated phenomenon.

The apparatus for studying the secondary electron emission of indium and tin in various states of aggregation constitutes a quasi-spherical condenser with an antidynatron (screen) grid [7]. The melting of the target is effected by passing a steady current through a spiral wound on a porcelain cylinder surrounding the test material.

In such investigations the vacuum conditions are of particular importance as well as the cleaning of the target surface from various kinds of contamination. In order to accelerate the measurements we used an oscillograph technique [6]. Clean target surfaces were obtained by means of a magnet-controlled attachment serving for the removal of surface-layer contamination in high vacuum.

In order to reduce possible errors associated with the measurement of the primary and secondary currents, we used an attachment to the oscillograph comprising two tubes and a switching relay. The use of this attachment enabled us to produce and measure pulses of primary and secondary currents (9 to 10 A) on the oscillograph screen. At the same time as measuring the magnitudes of these pulses for various primary-electron energies, we plotted the secondary-electron emission coefficient (σ) as a function of the primary electron energy E_p, i.e., $\sigma = f(E_p)$.

Results and Discussion. Using the method just described, we obtained the dependence of σ on E_p for solid and liquid indium not subjected to special purification (Fig. 1a). The secondary electron-emission coefficient increases substantially in the liquid state at low energies; it falls at high energies in the solid state. The value of σ for the liquid state reaches its maximum at energies smaller than those corresponding to the maximum for the solid state.

The relation between the secondary-emission coefficient and the primary electron energy for clean (repeatedly-purified) indium in the liquid and solid states is shown in Fig. 1B. The $\sigma = f(E_p)$ relationship has a maximum at almost the same primary-electron energy (550 eV) for both states, the difference between the σ of the solid and liquid states not exceeding 2 to 3%, which is within the limits of experimental error. This difference remains unaltered as the

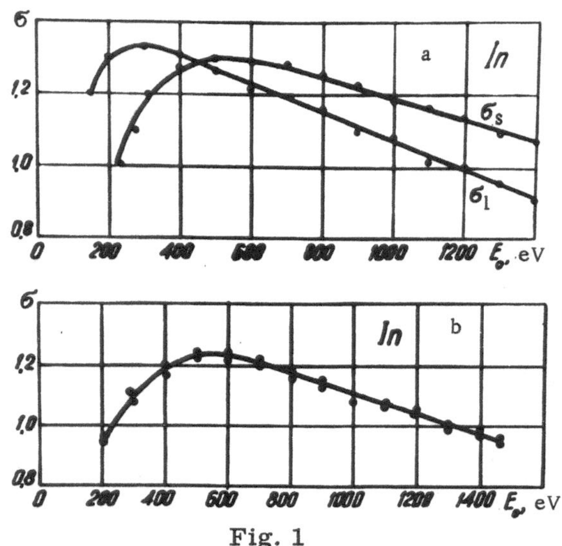

Fig. 1

primary energy increases (Fig. 1b). However, if the sample is kept for a long time in a vacuum of 10^{-6} mm Hg after cleaning in the manner indicated, there is a tendency for the value of σ in the liquid state to increase relative to that in the solid.

This increase is the greater, the longer the metal remains in a vacuum of 10^{-6} mm Hg after purification.

We see from the foregoing that the coefficient of secondary-electron emission varies with the surface conditions of the test material. The value obtained for the liquid, unpurified state is greater than that obtained for the solid state at low primary-electron energies. As E_p increases, this difference changes sign, i.e., for high E_p, $\sigma_l < \sigma_s$ (Fig. 1a). This behavior of the secondary electron-emission coefficient of indium in the liquid and solid states is always observed when the target surface is covered with an oxide film or a film of adsorbed gas; these may be observed visually through a special aperture in the receiving part of the apparatus.

The fact that the value of σ obtained for indium remaining for a long time in a vacuum of 10^{-6} mm Hg falls on passing from the liquid to the solid state, even after purification, is apparently a result of the effects of adsorbed films, comprising residual gas atoms. This assertion is supported by the fact that the intensity of the characteristic-loss peaks diminishes on increasing the period spent in a vacuum of 10^{-6} mm Hg after target cleaning [6-9], the reduction being the more rapid, the higher the residual-gas pressure.

On the basis of the foregoing data and also existing results, we may conclude that the coefficient of secondary-electron emission of indium and tin remains unaltered on passing from one state of aggregation into another. The increase or decrease in the coefficient of secondary-electron emission sometimes observed on melting or freezing is a consequence of a change in the conditions on the target surface, or the appearance of foreign compounds; it is independent of the electrophysical properties of the metal.

References

1. P. M. Morozov, Zh. Éksp. Teor. Fiz., 11:411 (1941).
2. V. G. Bol'shov and V. K. Seleznev, Zh. Tekh. Fiz., 14:1657 (1956).
3. V. G. Bol'shov and V. V. Zarubin, Fiz. Tverd. Tela, 1:462 (1959).
4. J. Brophy, Phys. Rev., 83:534 (1951).
5. Yu. M. Kushnir, V. I. Milyutin, and V. P. Goncharov, Zh. Tekhn. Fiz., 9:1589 (1959).
6. U. A. Arifov and A. Kh. Kasymov, Radiotekh. i Élektronika, 8:138 (1963).
7. U. A. Arifov and A. Kh. Kasymov, Dokl. Akad. Nauk SSSR, 158:82 (1954).
8. U. A. Arifov and A. Kh. Kasymov, Izv. Akad. Nauk Uzbek SSR, Ser. Fiz.-Mat. Nauk, No. 4, p. 92 (1963).
9. U. A. Arifov and A. Kh. Kasymov, Izv. Akad. Nauk Uzbek SSR, Ser. Fiz.-Mat. Nauk, No. 2, p. 23 (1965).

ENERGY DISTRIBUTION OF SECONDARY-ELECTRON EMISSION

U. A. Arifov and A. Kh. Kasymov

Many papers relating to the theoretical and experimental study of inelastic scattering, and particularly the associated characteristic electron energy losses in solid materials, have appeared in recent years [1-4]. An elucidation of the nature of the characteristic losses would provide us with more precise information as to the electron structure of solids, the processes involved in electron microscopy, cathodo-luminescence, semiconductor electronics, the passage of radiation through solids, and so forth.

Until recently, characteristic losses were mainly studied by firing primary electrons of high energy through thin foils; this method failed to reveal the complete spectrum. A study of the characteristic losses over the whole energy spectrum of the secondary electrons should throw fresh light on the relationship between this spectrum and the type of metal involved, its electron structure, and its atomic number (z). The relation between z and the energy spectrum of the secondary electrons has been studied in a number of cases, but no unified opinion has emerged. Owing to the practical difficulties of studying the angular dependence of the characteristic losses by the transmission method, results have only been obtained for small scattering angles; there is no information as to the dependence of the complete energy spectrum of the secondary electrons (and its individual components) on the angle of incidence of the primary electrons over a wide range of the latter.

Only one attempt has been made at studying the characteristic losses in relation to the state of aggregation, and this was carried out by the insufficiently accurate "transmission" method. In order to determine the true characteristic-loss spectrum, the spectrum of the substrate should also be known.

In this paper we shall analyze the secondary-electron energy spectrum by a method based on the use of spherical and cylindrical condensers. In order to accelerate the process and incorporate a degree of automation, we decided on the double-modulation oscillograph technique developed by U. A. Arifov, A. Kh. Ayukhanov, and S. V. Starodubtsev [5]. The combination of these methods unites the high resolving power of the cylindrical condenser, the great light-gathering power (transmission) of the spherical condenser, and the high-speed recording of the double-modulation procedure in a single piece of apparatus. Details of the measuring method and experimental apparatus appear elsewhere [6, 7].

The experiments confirmed the extreme importance of cleaning the target surface when studying the energy spectrum of the secondary electrons. In the case of inadequately-cleaned targets, the spectrum consists of slow secondary and inelastically-reflected electrons; hardly

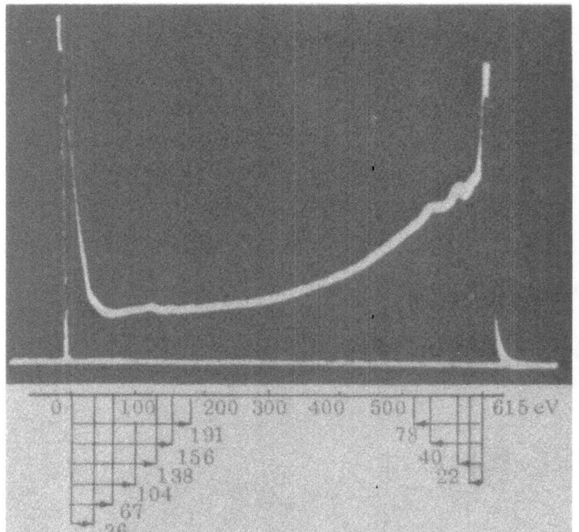

Fig. 1. Oscillogram of the secondary-electron energy spectrum for an Mo target with a primary-electron energy of 615 eV.

any elastically-reflected electrons occur. After adequate target cleaning at high temperatures, maxima appear in the low-energy range, together with peaks representing the characteristic secondary-electron loss spectra; for this reason, special attention was always devoted to cleaning the target surface. The duration of the heat treatment was determined by the reproducibility of the results characterizing the particular material under test; in the majority of cases it lasted several hours. The targets were also heated for one or two seconds immediately before each measurement, the oscillograms being photographed at a temperature close to the melting point of the test material.

We studied the secondary-electron energy distribution of a number of metals (Ti, Fe, Ni, Zr, Mo, In, Sn, Ta, W) widely employed in electrical vacuum apparatus. These metals, furthermore, constituted elements of the transition group, incomplete d or f layers appearing in their energy bands. A study of the energy distribution of the secondary electrons in these metals enabled us to relate the complete secondary-electron spectrum and the characteristic-loss peaks to the nature and electron structure of the test material. The experiments showed that the secondary-electron energy spectrum and the fine structure of the peaks were dependent on the internal structure of the substance in question. There was a specific relationship between the secondary-electron spectra of metals with a similar electron structure. For example, the low- and high-energy parts of the complete secondary-electron spectrum of Fe and Ni are similar in form and contain steps in the low-energy region (at 39 and 52 eV for iron and nickel respectively), as well as characteristic loss peaks.

The energy distributions of the secondary electrons in Ti and Zr are also similar in form. However, in contrast to the steps observed in the Fe and Ni spectra, these have maxima in the low-energy region. The secondary-electron spectra of indium and tin showed no maxima in the low-energy region, and the characteristic energy-loss peaks were comparatively small.

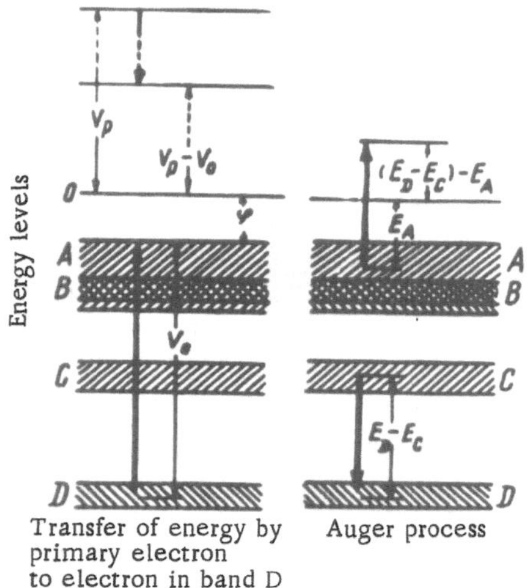

Fig. 2. Simplified description of the Auger process.

In considering the secondary-electron energy-distribution curves, we encounter the question: Which groups of electrons affect the development of the individual maxima on these curves? The end maximum on the right constitutes electrons elastically reflected by the emitter, the first maximum on the left is due mainly to secondary, and perhaps partly to inelastically-reflected electrons (Fig. 1).

1. Auger Electrons. The maxima (and bends) in the low-energy part of the secondary-electron spectrum (left-hand side of the spectrum) obtained for a number of metals (Mo, W, Ta, Ni, Ti and Zr) constitute electrons ejected as a result of the absorption of a certain proportion of the energy transferred by the incident electrons.

The maxima in the low-energy part of the secondary-electron spectrum with energies of 138 and 156 eV for Mo, 158 and 266 eV for W, 24, 30, 58, 123, and 145 eV for Ta,

148 eV for Ni, 52 and 145 eV for Ti, and 29, 96, 122, and 159 eV for Zr may apparently be attributed to metal electrons emitted as a result of the Auger process. In this case two electrons take part in the Auger process (Fig. 2). The first electron, initially situated in the C band, occupies a vacant place in the D band created as a result of the incidence of a primary electron. This releases energy $E_D - E_C$, leading to the detachment of a second electron initially situated, for example, in the A band. The second electron is emitted from the metal, losing energy E_A, and this is observed in the form of a maximum in the low-energy region of the secondary-electron spectrum. The probability of such a process is almost independent of the angle of incidence of the primary electrons; it depends on the primary-electron energy and intensity, as confirmed by measuring the intensity of such maxima as a function of the angle of incidence of the primary electrons.

2. Diffraction Effects. It is well known that diffraction phenomena obey the Wulff–Bragg formula

$$2d \sin \alpha = n\lambda$$

where d is the lattice constant, λ is the wavelength, n is the order of diffraction, and α is the glancing angle at which the diffraction is observed.

The development of peaks in the high-energy region of the secondary-electron spectrum is hard to explain as being due to diffraction effects, since the position of these peaks would be determined exclusively by the absolute energy of the primary electrons.

Our results indicate that the position of the peaks neighboring the maximum corresponding to the elastically-reflected electrons is independent of the energy of the incident electrons. On varying the energy by several hundreds of electron volts the position of the peaks relative to the zero position moves by the same amount. This is never observed in the case of diffraction.

3. Interaction of the Primary Electron with the Electron Gas of the Material as a Whole. Electrons which have lost small, discrete amounts of energy have been observed in a number of investigations. This may be explained on the basis of the Pines and Bohm theory [3] by considering that a fast primary electron interacts with the electrons of the sample material as if with a plasma. Comparison of the experimental results obtained with the theoretical values [3] shows that many of the characteristic losses actually observed can hardly be explained as being due to the interaction of the primary electrons with an electron gas.

4. Inelastic Scattering of Primary Electrons. If we are to explain the phenomena observed as being due to the inelastic scattering of the primary electrons in solids, the secondary electron emission should contain a group of inelastically-reflected electrons which have lost a specific amount of energy, characteristic of the particular material. The existence of such energy losses by the primary electrons should, furthermore, lead to the appearance, not only of inelastically-reflected electrons but also (among the slow secondary electrons) of groups of electrons with an energy equal to that released by the primary electron. Hence the maxima under consideration may correspond to both secondary and inelastically-reflected electrons. If individual maxima are associated with secondary electrons only, then the position of these maxima should be independent of the energy of the primary electrons and should be constant relative to the zero energy of the electrons. If, however, the maxima constitute the result of the inelastic reflection of primary electrons which have lost a specific amount of energy, then their position should be constant relative to the peak of the elastically-reflected electrons.

Thus if inelastically-scattered and secondary electrons both participate in the creation of the intermediate maxima, there should in general be two systems of peaks, each of which will maintain a constant position relative to either the origin of coordinates or the elastically-scattered electron peak as the primary-electron energy varies.

The existence of continuous and discrete spectra is not surprising from the foregoing point of view, as there are two possible mechanisms of primary-electron scattering in metals, namely, scattering by either the free or the bound electrons of the metal.

A comparison of our own characteristic-loss data with earlier published results indicates that the results of the oscillograph method of recording the complete secondary-electron spectrum agree satisfactorily with the corresponding characteristic-loss peaks derived in other ways.

The steps observed [8] in the secondary-electron spectra of Fe and Ni result from the incomplete energy bands of these metals. This explanation may be regarded as plausible because, firstly, among all the metals employed such steps only appear for Fe and Ni, secondly, the height of the steps depends on the width of the incomplete band in the metal, and, thirdly, the number of steps in the secondary-electron spectrum of Fe depends on the presence of a certain amount of Ni in its composition. These steps may be ascribed to a group of electrons formed by electrons belonging to the metal, situated in the filled (occupied) part of the d band; they result from the interaction of the primary electrons with these metal electrons. It is possible that electrons having energies smaller than the energy of the step are at least partly secondary electrons while electrons having higher energies than that of the step are inelastically reflected. This point of view is supported by the fact that the dependence of the height of this step on the angle of incidence of the primary electrons is similar to that of the slow secondary electrons. The relationship between the height of the step and the angle of incidence of the primary electrons has a maximum at incident angles very similar to the reflection angle. The intensity of the maxima in the low-energy part of the secondary-electron spectrum is almost independent on the incident angle of the primary electrons. Thus the chief process determining the characteristic losses in the metals under consideration lies in the interband transitions of individual electrons under the influence of the primary beam.

References

1. A. R. Shul'man (ed.), Characteristic Energy Losses of Electrons in Solids, Moscow (1958).
2. N. Fridman, Uspekhi Fiz. Nauk, 62:427 (1957).
3. D. Pines, Uspekhi Fiz. Nauk, 62:399 (1957).
4. A. Ya. Vyatskin, Zh. Tekhn. Fiz., 28:2217, 2455 (1958).
5. U. A. Arifov, A. Kh. Ayukhanov, and S. V. Starodubtsev, Zh. Tekhn. Fiz., 26:714 (1954).
6. U. A. Arifov and A. Kh. Kasymov, Radiotekhnia i Élektronika, 8:138 (1963).
7. U. A. Arifov and A. Kh. Kasymov, Izv. Akad. Nauk Uzbek SSR, Ser. Fiz.-Mat. Nauk, 2:23 (1965).
8. U. A. Arifov and A. Kh. Kasymov, Izv. Akad. Nauk Uzbek SSR, Ser. Fiz.-Mat. Nauk, 4:92 (1963).

EFFECT OF MULTIPLE SCATTERING AND THERMAL VIBRATIONS ON THE REFLECTION OF IONS FROM THE FACE OF A SINGLE CRYSTAL

É. S. Parilis, N. Yu. Turaev, and V. M. Kivilis

The characteristics of the two-fold collision peaks in the energy spectrum of ions reflected from the face of a single crystal in a specified direction were calculated in earlier papers [1-3]. Later [4] a model based on scattering by two neighboring atoms was used in order to show that, as a result of the mutual screening of these atoms, both minimum and maximum angles of reflection should exist. However, the two-atom scattering model is imprecise. For glancing incidence on the face of a single crystal in a plane passing through a direction of close packing, the ion suffers a number of successive collisions with series of atoms forming chains in the crystal lattice. The parameters of each successive collision are determined by those of all the preceding collisions. A direct calculation of collisions of this kind for chains consisting of immovable atoms gives an unexpected result [5]. The energy spectrum of the ions reflected in a specific direction as a result of five or ten collisions also contains the peaks of "single" and "double" collisions, the single ones constituting the result of a single strong and a large number of weak collisions, while the double ones consist of two successive relatively strong and a number of weak collisions.

However, both the position of these peaks and the distance between them differ sharply from those calculated on the two-atom model; the angular limits of the reflected beam are narrow and contract more strongly for more glancing incidence, becoming compressed around the direction of specular reflection.

In order to study the effect of the thermal vibrations of the lattice atoms on the foregoing phenomena, the Einstein model of independent harmonic oscillators is quite inapplicable. In the present case, it is essential to allow for the correlation between the vibrations of neighboring atoms forming the chain. The situation is analogous to that encountered by Nelson, Tompson, and Montgomery when studying the effect of thermal vibrations on focused collisions [6]. We therefore considered the chains in a monatomic, face-centered lattice as one-dimensional chains, the phonon spectrum of these containing all wavelengths above the minimum wavelength $\lambda = 2d$ in the Debye theory, corresponding to the maximum (and strongest) frequency ω_{max}. (To be more accurate, there are two transverse vibrations and one longitudinal vibration for each frequency, the period of these being $T > 10^{-11}$ sec.) A fast ion (traveling at 10^6 to 10^8 cm/sec) approaches and leaves a bent, rigid chain in a period of 10^{-13} to 10^{-15} sec (Fig. 1).

The possibility of observing any particular peak experimentally is determined by its half-width, which depends considerably on the thermal vibrations of the lattice atoms around their

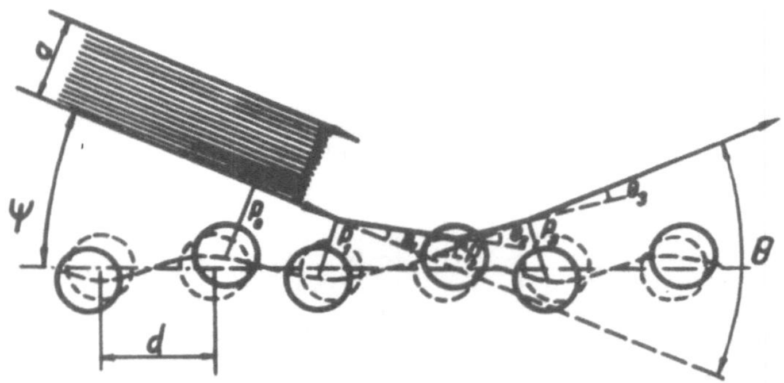

Fig. 1. Scattering of ions by a one-dimensional isolated chain of atoms (schematic).

equilibrium positions. Assuming that the atoms vibrate perpendicularly to the line joining them, we may calculate the degree of blurring (spread) of the secondary peaks in the energy spectrum in relation to the temperature of the lattice.

The deviation of an atom from the equilibrium position may be sought in the form [8]

$$(\bar{s}_0)^2 = \frac{1}{m_2 N} \int \frac{\varepsilon(\omega, T)}{\omega^2} Z(\omega)\, d\omega$$

where $Z(\omega)$ is the frequency-distribution function in the solid,
$\varepsilon(\omega, T)$ is the thermal energy of a linear oscillator,
N is the number of atoms in 1 cm^3 of the solid.

The deviation in any one direction is characterized by the value of $\gamma = (\bar{s}_0)^2/3$, and the root of this expression characterizes the linear "spread" in the position of the atom.

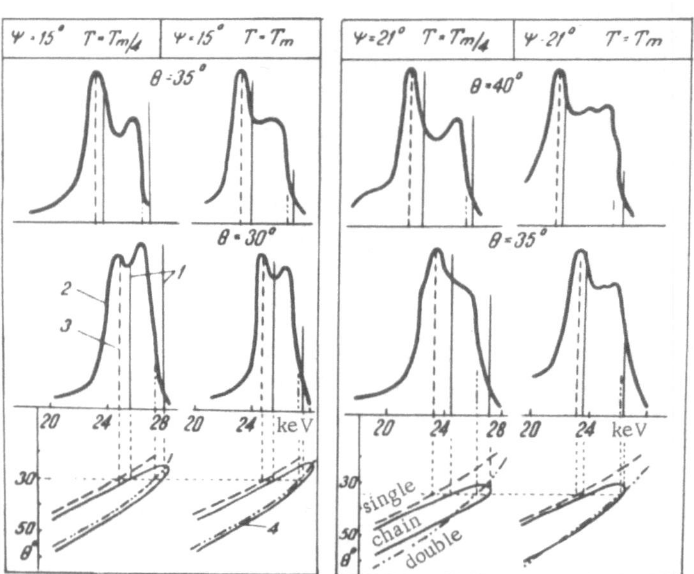

Fig. 2. Energy spectra of Ar^+ ions reflected from a one-dimensional ⟨110⟩ chain in a nickel single crystal. 1) Theoretical data; 2) experimental [7]; 3) position of the single peak [7]; 4) E (Θ) relationship, similar curves for purely single and double collisions shown alongside for comparison.

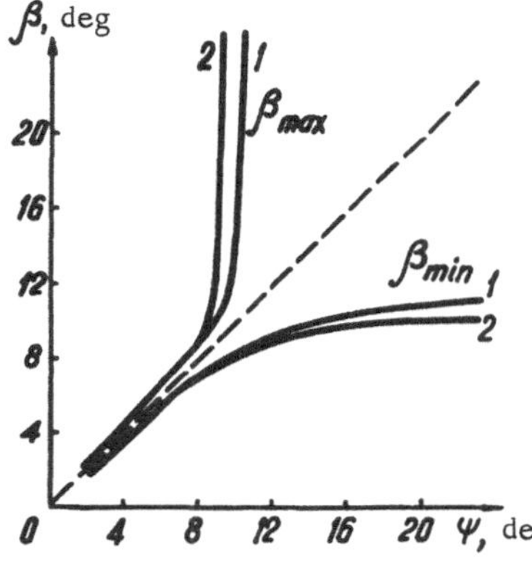

Fig. 3. Limiting angles of escape in relation to the angle of incidence for T = $T_m/4$ (1) and T = T_m (2).

For very high and very low temperatures

$$\delta(T) = \left[\frac{1}{3}(\overline{s_0})^2\right]^{1/2} \approx 10^{-1} \cdot d \begin{cases} \sqrt{T/T_m} & \text{for } T \gg \Theta_D \\ \sqrt{\Theta_D/4T_m} & \text{for } T \sim 0 \end{cases}$$

where Θ_D is the Debye temperature of the solid in question,
T_m is the melting point,
T is the absolute temperature,
d is the distance between the closest atoms in the crystal.

The amplitude of the waves increases as $\sqrt{T}$ and is equal to 0.1 d near the melting point of the metal and 0.05 d at T = $T_m/4$.

Our calculations of the multiple scattering of 30-keV of Ar^+ atoms from the (110) chain of nickel targets, using the Firsov potential and allowing for both elastic and inelastic energy losses with all possible collision parameters (incorporating averaging over the phonon spectrum), were carried out in the same way as in the earlier treatment [5].

The aiming parameter of a secondary collision is determined by that of the previous collision p_i and the scattering angles Θ^i in the following manner:

$$p_{i+1}[E_i(\Theta_{i+1})] = \{p_i[E_{i-1}(\Theta_{i-1}), \Theta_i] - d\sin(\psi - \sum\Theta_i) + \\ + (-1)^i\, 2\delta T[\cos(\psi - \sum\Theta_i) - \sin(\psi - \sum\Theta_i)]\}$$

Results typifying these calculations are presented in Fig. 2; here $\Theta = \Sigma\,\Theta_i$ is the total scattering angle, while $E = E_0 - \sum_i (\Delta E_{el} + \Delta E_{inel})_i$. The calculated peaks for Θ = 30°, 35°, and 40° may be compared with the experimental values [7]. The contraction of the reflected beam due to multiple scattering and heating is indicated in Fig. 3. The double-scattering peaks diminish on heating and their half-width increases, in agreement with earlier considerations [5]. The relative intensities of the twofold scattering peaks obtained from the chain calculations agree with the experimental values better than the intensities derived from the two-atom model. We see from Fig. 2 that the scattering by the chain produces a small but temperature-dependent displacement of both single and twofold peaks and also brings them closer together. On further increasing the resolving power of the apparatus, careful measurement of the position of the peaks on the energy scale may lead to the experimental observation of this displacement and enable us to relate it to the thermal vibrations; this should offer a unique opportunity of using the reflection of fast ions in order to carry out a direct investigation into the correlation between the vibrations of chain-forming atoms.

We consider that, despite its obviously simplified nature, the model here described provides a valid interpretation of the true nature of the scattering of fast ions from the face of a single crystal at small angles of incidence; it may accordingly be used to discuss both known experimental results and also those still expected to be revealed.

References

1. É. S. Parilis and N. Yu. Turaev, Dokl. Akad. Nauk Uzbek SSR, 12:16 (1964).
2. E. S. Mashkova, V. A. Molchanov, É. S. Parilis, and N. Turaev, Phys. Letters, 18:7 (1965).
3. E. S. Mashkova, V. A. Molchanov, É. S. Parilis, and N. Yu. Turaev, Dokl. Akad. Nauk SSSR, 166:330 (1966).
4. É. Parilis, Proc. Seventh Internat. Conf. on Ionization Phenomena in Gases, Belgrade (1965).
5. V. M. Kivilis, É. S. Parilis, and N. Yu. Turaev, Dokl. Akad. Nauk SSSR, 173:1983 (1967).
6. R. S. Nelson, M. E. Tompson, and H. Montgomery, Phil. Mag., 7:1385 (1962).
7. É. S. Mashkova, V. A. Molchanov, and V. Sozka, Phys. Stat. Sol., 19:425 (1967).
8. I. Leibfried, Handbuch der Physik, Vol. 7 (1955), p. 262.

AUGER IONIZATION OF ATOMS BY MUTIPLY-CHARGED IONS. MODEL OF TWO COULOMB CENTERS

L. M. Kishinevskii and É. S. Parilis

When an n-charged ion B^{n+} collides with an atom A, a process similar to the autoionization of an excited atom [1] may occur, the $A + B^{n+}$ system relaxing rapidly by way of the Auger effect [2]. This phenomenon is analogous to potential ion–electron emission [3] and Penning ionization [4].

The Auger-ionization process may occur in at least three ways [5]; directly (I), with preliminary one-electron charge exchange (II), and with preliminary two-electron charge exchange (III); excited states of the ion result from these processes.

The ionization cross section $\sigma_k(v)$ for the k-th channel (k = I, II, III) corresponding to a collision with an ion traveling at a velocity v is related to the probability $P_k(\rho, v)$ of ionization taking place as the ion travels along a trajectory with a collision parameter ρ in the following manner [5]:

$$\sigma_k = 2\pi \int_0^\infty P_k(\rho, v)\rho d\rho \tag{1}$$

where

$$P_k(\rho, v) = w_k(1 - \omega_k) \tag{2}$$

if Auger ionization is effective within the region between the points of intersection of the terms R_i in which charge-exchange takes place; w_k is the probability of falling into the k-th channel;

$$\omega_k = \exp\left[-\int_{-t_i}^{t_i} W_k[R(t)]\,dt\right] \tag{3}$$

is the weakening of the flux of incident ions due to Auger ionization, and W_k is the probability of Auger ionization in unit time (for fixed nuclei).

Wave Functions and Energies of the Electrons. The general aspects of the Auger ionization of atoms under the influence of multiply-charged ions were considered earlier [5] on the basis of certain relationships, valid for large interatomic distances R. In order to derive some numerical estimates over the whole range of interatomic distances $0 \leq R \leq \infty$, we shall here consider Auger ionization for the case in which the ion B^{n+} is represented by a Coulomb center, $Z_B = n$, and the atom by $Z_A = 2$, with two electrons (helium atom) [6]. The perturbation arises from the Coulomb interaction of these electrons $1/r_{12}$ (here and subsequently we shall use atomic units: $e^2 = m_e = \hbar = 1$). Hence the single-electron wave functions may be obtained by solving the Schrödinger equation for two centers:

$$\left(-\frac{1}{2}\Delta + \frac{Z_A}{r_A} + \frac{Z_B}{r_B}\right)\psi = E\psi \tag{4}$$

$$\psi(\mu, \nu, \varphi) = M(\mu) N(\nu) e^{im\varphi} \tag{5}$$

Here $\mu = \frac{r_A + r_B}{R}$; $\nu = \frac{r_A - r_B}{R}$; φ are elliptic coordinates.

The solution of Eq. (4) has been studied by a number of authors [7-12]. For interatomic distances R not exceeding a few units, we used the solutions of Bates and Carson [10] for $Z_B = 2Z_A$, while for other values of Z_B we used the method of solution indicated by the same authors. The behavior of $\psi(\mu, \nu, \varphi)$ in the neighborhood of each of the nuclei was considered for $R \gg 1$ by Gershtein and Krivchenkov [12]. In this case it is convenient to use an expansion in the form

$$M(\mu) = (\mu^2 - 1)^{|m|/2} e^{-p\mu} \sum_{n_1=0}^{\infty} c_{n_1} F[-n_{n1}, |m|+1, 2p(\mu - 1)] \tag{6}$$

$$N_{\pm}(\nu) = (\nu^2 - 1)^{|m|/2} e^{\pm p\nu} \sum_{n_2=0}^{\infty} g_{n_2} F[(\beta - n_2), |m|+1, 2p(1 \mp \nu)] \tag{7}$$

where $p^2 = -(R^2E)/2$. An electron localized in the A nucleus is described for $R \to \infty$ by the function $N_+(\nu)|_{\beta=0}$ or the function $N_-(\nu)|_{\beta=-b/2p}$, where $b = (Z_A - Z_B)R$. We note that the expansion of $N_{\pm}(\nu)|_{\beta=0}$ only converges rapidly in the neighborhood of the particular nucleus in which the electron is localized, while the expansion of $N_{\pm}(\nu)|_{\beta=\pm b/2p}$ does so for all possible values of ν $(-1 \leq \nu \leq 1)$. This circumstance was not mentioned in the earlier treatment [12]; it yields the following relationship between the pairs of linearly-dependent solutions in question:

$$D_{\pm} = \frac{N_{\mp}(\nu)|_{\beta=0}}{N_{\pm}(\nu)|_{\beta=\pm b/2p}} \tag{8}$$

The behavior of the wave functions of the continuous spectrum for $R \gg 1$ is also determined by the solutions (6) and (7), which, in the neighborhood of each of the centers, depend only on the coordinates of that particular center. This enables us to use any complete set of single-centered functions as wave functions of the continuous spectrum, including the set of eigenfunctions of the continuous spectrum expressed in spherical coordinates around each of the centers. i.e., $\{\psi_E\} \equiv \{\psi_{Elm}\}_A$ and $\{\psi_{Elm}\}_B$.

The eigenvalues of Eq. (4) determine the terms of the quasi-molecule and the energies of the Auger electrons $E_k^0(R)$ in the zero approximation.

Initial and Final States of the System. Conversion into Excited States by Charge Exchange. In the initial state of the system the electrons occur in the state $(1s)^2$ in the atom A; in the quasi-molecule this corresponds to the state $(2p\sigma)^2$ for $1 < Z_B/Z_A \leq 2$, the state $(3d\sigma)^2$ for $2 < Z_B/Z_A \leq 3$, and so on [12]. Let us consider the case in which $Z_B = 3(Li^{3+})$ and $Z_B = 4(Be^{4+})$.

TABLE 1.

States with intersecting terms	R_i, at. units	$\Delta U_i(R_i)$, at. units	$\|F_0-F_i\|$, at. units	v_i, cm sec
$(2p\sigma)^2-(2p\sigma)(2s\sigma)$	$R_s=2.8$	0.05	0.5	$v_s=1.7\cdot10^6$
$(2p\sigma)^2-(2p\sigma)(3d\sigma)$	$R_d=2.64$	0.48	0.9	$v_d=8.8\cdot10^7$
$(2p\sigma)^2-(2p\sigma)(2p\pi)$	2.6	0	—	0
$(2p\sigma)(3d\sigma)-(2p\sigma)(2p\pi)$	2.8	0	—	0
$(2p\sigma)(3d\sigma)-(2p\sigma)(2s\sigma)$	$R_{sd}=2.47$	0.05	0.3	$v_{sd}=2.9\cdot10^6$

As the atom A approaches the nucleus B, the term representing the initial state intersects the terms corresponding to the transition of one of the electrons into an excited state in the ion with principal quantum number $n_B = 2$, which becomes degenerate as $R \to \infty$. In the quasi-molecule the states $(2p\sigma)(2s\sigma)$, $(2p\sigma)(2p\pi)$, $(2p\sigma)(3d\sigma)$ correspond to this [12].

In Auger ionization, one of the electrons passes into the ground state of the ion (1s) and the other into the continuous spectrum. The probability that the electron will not pass into the other atom when the terms intersect is given by the Landau–Zener formula [13, 14]:

$$p_i = \exp\left[-\frac{v_i}{v_{R_i}}\right], \qquad v_i = \frac{\pi}{2}\frac{[\Delta U_i(R_i)]^2}{|F_0 - F_i|} \tag{9}$$

($i = 1, 2$ for one- and two-electron charge exchange). Here v_{R_i} is the radial velocity, $\Delta U_i(R_i)$ is the separation of the terms at the point of intersection R_i, and F_0 and F_i are the strengths of the terms.

The charge exchange between Li^{3+} and He reflects the intersection of the terms for $R_1 = 8.5$; $\Delta U_1 = 1.6\cdot10^{-3}$; $v_1 = 3.5\cdot10^4$ cm/sec. Transitions between excited states in the Be^{4+}–He system involve the intersection of the terms categorized in Table 1.

In this system, two-electron charge exchange corresponds to the intersection of terms with $R_2 = 9.3 \gg 1$ so that $v_2 < 10^4$ cm/sec, i.e., for practical velocities this kind of charge exchange fails to occur. The probabilities that, after a single passage of the points R_s, R_d, R_{sd}, the system will be in the states $(2p\sigma)^2$, $(2p\sigma)(2s\sigma)$, $(2p\sigma)(3d\sigma)$ are described by the relations:

$$w_{(2p\sigma)^2} = p_s p_d \tag{10}$$

$$w_{(2p\sigma)(2s\sigma)} = p_s(1-p_d)(1-p_{sd}) + (1-p_s)p_{sd} \tag{11}$$

$$w_{(2p\sigma)(3d\sigma)} = p_s(1-p_d)p_{sd} + (1-p_s)(1-p_{sd}) \tag{12}$$

Probabilities of Auger Ionization for Fixed Nuclei. The probabilities of an Augur transition via channel I into the states $\lambda s\sigma$ and $\lambda d\sigma$ for $R < 1$ are at least two orders of magnitude greater than the probabilities of transitions into other states of the continuous spectrum (here λ characterizes the energy of the Auger electron $E_I^{(0)} = 2\lambda^2/R^2$). In this region $W_I(R)$ may be represented by the following approximate relation:

$$W_I(R) = W_I|_{R=0}\cdot\frac{Z_{1s\sigma}^3 Z_{2p\sigma}^{10}}{(Z_A+Z_B)^{13}}\left[\frac{3(Z_A+Z_B)}{2Z_{1s\sigma}+Z_{2p\sigma}}\right]^{2t} \tag{13}$$

where

$$t = 8 \text{ for } l = 0, \; t = 7 \text{ for } l = 2$$

$$Z_{1s\sigma} = \sqrt{-2E_{1s\sigma}(R)}, \; Z_{2p\sigma} = \sqrt{-8E_{2p\sigma}(R)}$$

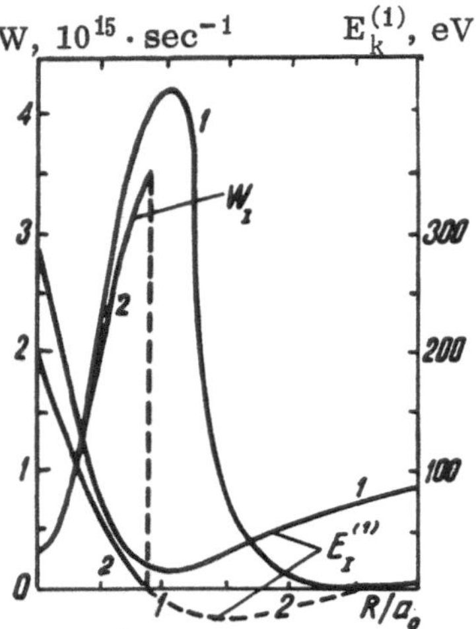

Fig. 1. The probabilities of Auger ionization $W_I(R)$ and energies of the Auger electrons $E_I^{(1)}(R)$: 1) $Z_B = 4$; 2) $Z_B = 3$.

In the region $R \gg 1$ the transition probability is exponentially small, since the "pullover" of the wave function from the atom to the ion falls exponentially with increasing R. According to Eq. (8) this "pullover" is given by the function

$$D_{2p\sigma} = \Gamma\left(\frac{Z_A - Z_B}{2p_{2p\sigma}} R\right) (4p_{2p\sigma})^{Z_B/Z_A} e^{-2p_{2p\sigma}} \tag{14}$$

$$p_{2p\sigma} = \frac{Z_A R}{2} + \frac{Z_B}{2Z_A} + 0(R^{-1}) \tag{15}$$

while the probability of ionization in channel I [5] is

$$W_I(R) \sim D^2_{2p\sigma} R^{-6} \tag{16}$$

Using the exact functions, we calculated the probabilities W_I for $Z_B = 4$ at the points R = 0.5, 1.5, 2, and 2.5 and for $Z_B = 3$ at R = 0.5. For the average interatomic distances there is a range in which $E_I^{(0)} = 2E_{2p\sigma} - E_{1s\sigma} < 0$, i.e., Auger ionization is impossible. Allowance for the interaction of the electrons increases the energy of the Auger electrons; however, this cannot be used as a perturbation producing an Auger transition. The more complicated form of the perturbation greatly impedes calculations even in the case of the Auger ionization of an atom [16]. For this reason the transition probabilities calculated by the method described were extended in the direction of large R from the left and in the direction of small R from the right, i.e., the intermediate region was obtained by bilateral interpolation.

The total probabilities of Auger ionization for the $(2p\sigma)^2$ state considered are presented in Fig. 1. For comparison, the energies of the Auger electrons are also given. We notice the natural correlation between the Auger-ionization probability and the energy carried away by the electron: The maximum of the former coincides with the minimum of the latter.

It was shown earlier [5] that Auger ionization in channel II involves two possibilities: 1) the escape of the second atomic electron into the continuum, and 2) the emission of an electron from the excited ion. The transition probability

$$W_{II}(R) = 2\pi |V_{II}^{(1)} + V_{II}^{(2)}|^2 g_f \tag{17}$$

In the first case the matrix element of the transition is determined by dipole–dipole interaction [17-19]:

$$V_{II,m}^{(1)} = \left\langle \psi_{A,000}(1)\, \psi_{B*,n_1n_2m}(2) \left| \frac{1}{r_{12}} \right| \psi_{A,Elm}(1)\, \psi_{B,000}(2) \right\rangle \tag{18}$$

For two Coulomb centers

$$V_{II,m}^{(1)} = c_m \frac{2^{12}}{3^{11/2}} \left(Z_B Z_A^2 R^3\right)^{-1} \left(1 + \frac{2E_{II}}{Z_A^2}\right)^{-5/2} \exp\left[-\frac{Z_A \sqrt{2}}{\sqrt{E_{II}}} \arctan \frac{\sqrt{2E_{II}}}{Z_A}\right] \tag{19}$$

Here $c_0 = 1$; $c_1 = 2^{-0.5}$. In the case $Z_B = 4$, $V_{II,0}^{(1)} \approx 0.015\,R^{-3}$; $V_{II,1}^{(1)} \approx 0.021\,R^{-3}$; at interatomic distances $R \sim 1$ these transitions make a contribution to the total probability W_{II} equal to $\sim 10^{-3}$ (atomic time units)$^{-1}$ $\sim 10^{14}$ sec^{-1}.

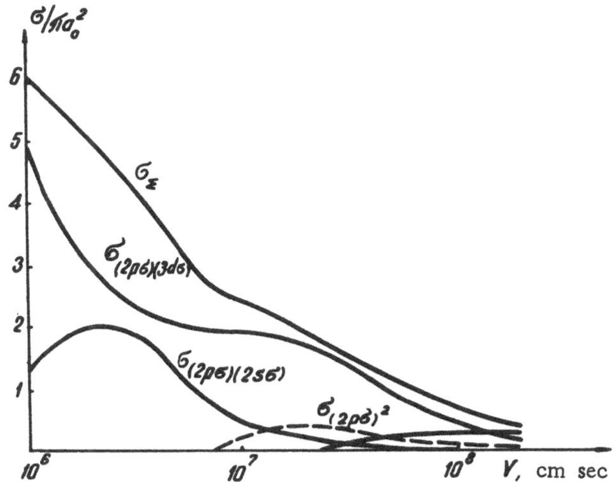

Fig. 2. Auger-ionization cross sections. Continuous lines: $Z_B = 4$; broken lines: $Z_B = 3$.

In the second case the matrix element of the transition

$$V_{II}^{(2)} = < \psi_{A,000}(1)\,\psi_{B^*,\,n_1n_2m}(2)\left|\frac{1}{r_{12}}\right|\psi_{B,Elm}(1)\,\psi_{B,000}(2) > D(R) \tag{20}$$

is nonzero in the case of $l = 0, 1, 2$ for transitions from the states $\psi_{B^*,100}$ and $\psi_{B^*,010}$ and in the case of $l = 1, 2$ for $\psi_{B^*,001}$.

At fairly large interatomic distances ($R > 10$), $V_{II}^{(1)}$ exceeds $V_{II}^{(2)}$. However, at distances of the order of the dimensions of the atom, D(R) falls off more slowly than R^{-3}, owing to the preexponential factor, and in the region of effective Auger ionization $V_{II}^{(1)} \ll V_{II}^{(2)}$, so that $W_{II} \sim D^2(R)$. For the transitions just considered this equals:

$$W_{II,\,100} \approx 0.026D^2;\quad W_{II,\,010} \approx 0.061D^2;\quad W_{II,\,001} \approx 0.009D^2$$

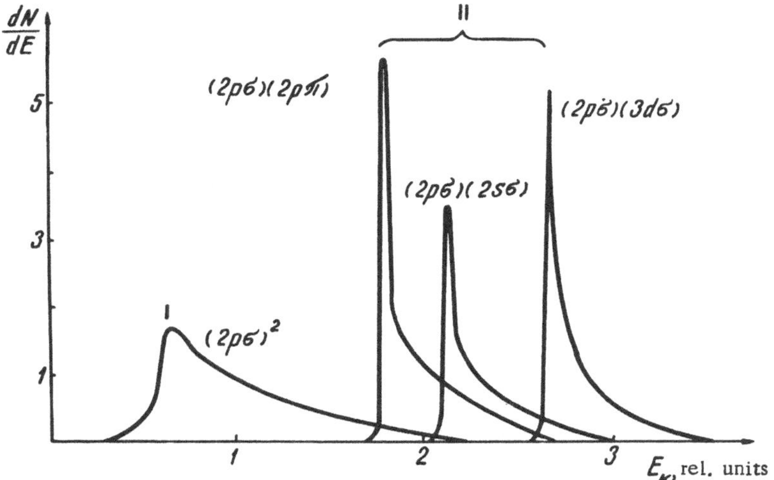

Fig. 3. Energy spectrum of the Auger electrons for Z_B = 4 (arbitrary units); I, II indicate the corresponding channels of the process.

Ionization Cross Sections; Energy Spectra of the Electrons. We calculated the Auger-ionization cross sections from relations (1) to (3). In the case of $Z_B = 3$, when the velocity $v_1 = 3.5 \cdot 10^4$ cm/sec, ionization in channel II is effective for velocities $v < 10^6$ cm/sec. Ionization in channel I from the initial state $(2p\sigma)^2$ only becomes possible after passing a certain critical velocity v_t, for which a range of interatomic distance in which the ionization is energetically possible ($R \leq 0.87$) is achieved. Estimates give $v_t = 7.6 \cdot 10^6$ cm/sec for the critical velocity (Fig. 2).

In the case of $Z_B = 4$ the threshold effect is absent, and ionization may occur for any approach velocities. The manner in which the cross sections $\sigma_{(2p\sigma)(2s\sigma)}(v)$ and $\sigma_{(2p\sigma)(3d\sigma)}(v)$ vary, the first having a maximum and the second varying little with velocity in the range $2 \cdot 10^6$ to $2 \cdot 10^7$ cm/sec, is determined by charge-exchange competition, associated with the quasi-intersection of the terms at the points R_s, R_d, R_{sd}. For velocities $v > v_d$ the Auger-ionization cross section in channel I exceeds the ionization cross section in channel II. Figure 2 also shows the total ionization cross section $\sigma_\Sigma = \sum_k \sigma_k$. This varies more smoothly with velocity than the partial cross sections. It is the total rather than the partial cross sections which are measured experimentally. However, by measuring the energy spectrum of the Auger electrons, the partial cross sections may also be estimated.

The energy spectra for the case of $Z_B = 4$ calculated on the basis of the relationships derived earlier [5] are plotted in Fig. 3. As indicated earlier (Table 1), owing to the absence of static interaction in the intersection of the $(2p\sigma)^2$ and $(2p\sigma)(2p\pi)$ terms, no transition occurs from the first of these to the second. However, the strong coupling of states differing only in magnetic quantum number should lead to transitions between such states by virtue of the rotation of the internuclear axis [20]. Figure 3 therefore incorporates the energy spectrum for this case also. Although the flat parts of the spectra overlap to a considerable extent, the division into groups of peaks belonging to different channels is perfectly sharp.

The maxima in the center are due to the vanishing of the derivative dE_k/dR; hence their position is independent of the velocity of the ion, while their heights should vary with velocity $\sim \sigma_k(v)$. With increasing velocity the peaks in channel II should diminish, while those in channel I should rise.

References

1. L. M. Kishinevskii and E. S. Parilis, Trans. of the Second All-Union Conference on Atomic and Electron Collisions, Uzhgorod (1962).
2. É. S. Parilis, The Auger Effect, Izd. Fan, Tashkent (1969).
3. U. A. Arifov, Interaction of Atomic Particles with Solid Surfaces, Izd. Nauka, Moscow (1968).
4. A. A. Kruithof and F. M. Penning, Physica, 4:430 (1937).
5. L. M. Kishinevskii and É. S. Parilis, Zh. Éksp. Teor. Fiz., 55:1932 (1968).
6. L. M. Kishinevskii and É. S. Parilis, Fifth Internat. Conf. on the Physics of Electron and Atomic Collisions, Leningrad (1967), p. 100.
7. E. A. Hylleraas, Zs. Phys., 71:739 (1931).
8. H. Bethe, Quantum Mechanics of Very Simple Systems [Russian translation], ONTI, Moscow-Leningrad (1935).
9. G. Jaffe, Z. Phys., 87:535 (1934).
10. D. R. Bates and T. R. Carson, Proc. Roy. Soc., A234:207 (1956).
11. M. Shimizu, J. Phys. Soc. Japan, 18:811 (1963).
12. S. A. Gershtein and V. D. Krivchenkov, Zh. Éksp. Teor. Fiz., 40:1491 (1961).
13. L. D. Landau, Phys. Zs. Sowjetunion, 2:46 (1932).
14. C. Zener, Proc. Roy. Soc., A137:696 (1932).

15. T. J. M. Boyd and B. L. Moiseiwitsch, Proc. Phys. Soc., A70:809 (1957).
16. B. H. Brandsen and A. Dalgarno, Proc. Phys. Soc., A66:904 (1953).
17. K. Katsuura, J. Chem. Phys., 42:3771 (1965).
18. B. M. Smirnov and O. B. Firsov, ZhÉTF, Pis. Red., 2:478 (1965).
19. T. Watanabe and K. Katsuura, J. Chem. Phys., 47:800 (1967).
20. Atomic and Molecular Processes, Izd. Mir, Moscow (1964), pp. 520, 530.

ION—CURRENT CHARACTERISTICS IN THE ELECTRON-BEAM ZONE-MELTING OF MOLYBDENUM

I. A. Brodskii, E. E. Petushkov, and S. M. Mikhailov

In this paper we shall be considering phenomena associated with the ionization of gases evolved from a molten zone of metal; we shall indicate a method of measuring the ion current and derive a relationship between the ion current and the quantity of evolved gases in a molybdenum zone-melting installation.

In view of the relatively high melting point of molybdenum, gases are evolved very energetically and in great quantities from the zone of molten metal. At the same time there is an intensive sputtering of the metal; it settles on the walls of the chamber and has considerable gettering properties. Thus the evolution of gas from the metal is accompanied by its intensive absorption in the getter layer, so that the quantities of gas evolved cannot be measured by direct methods [1]. It is accordingly of considerable interest to establish a relationship between the ion current and the gas evolution of the zone. For this purpose we used a zone-melting installation with a conical cathode unit. This kind of unit (Fig. 1a) ensures a stable mode of zone melting by virtue of the spatial decoupling of the electron and ion currents of the melting equipment and a great weakening of positive feedback effects.

The formation of the free zone of molten metal enables the molecular and ion beams emitted by the zone to fall directly on the collector (casing of the equipment).

Under good vacuum conditions, the surface of the molten metal is uniform and may be represented as a uniform emitter of positive ions, formed in the evaporation of impurity atoms and molecules. The growth of a single crystal in the course of zone melting tends to drive impurities from the inside of the zone to the surface. In this case the impurity atoms and molecules may be represented as atoms adsorbed on the surface, or "adatoms."

In view of the high temperature of the zone of molten molybdenum (2640°C), surface ionization of various impurity atoms and molecules occurs on the surface in the weak electric field of the system formed by the anode (the molten metal sample) and the collector (cap of the melting equipment). The atoms and molecules evaporating from the surface of the zone also undergo intensive shock ionization by virtue of the electron beam from the cathode unit. The thermal radiation of the zone and the stopping of the electron beam in the latter create additional factors producing volume ionization by dint of ultraviolet radiation and x rays.

The effect of all these ionizing factors is concentrated on a relatively small area of the zone of molten metal and in a small region of the neighboring space; the ion current is therefore very strong, and its magnitude and character are related to the removal of impurities from the zone. From the initial instant of zone melting, there is a steady flow of impurity atoms

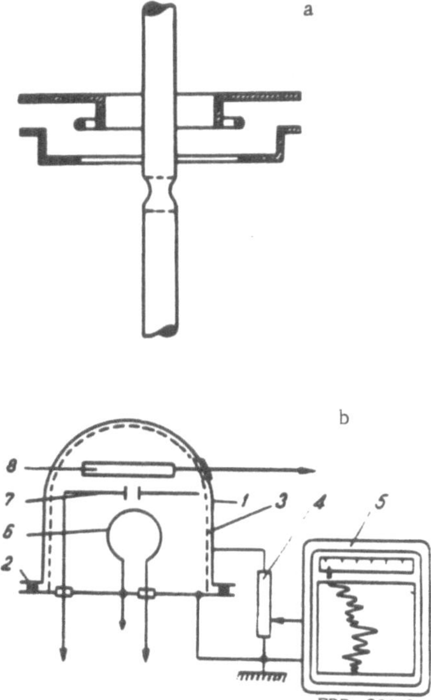

Fig. 1. Arrangement of the cathode unit (a) and recording of the ion current in the melting equipment (b): 1) metal cap (dome, cover); 2) rubber insulation (gasket); 3) grid; 4) variable resistance SP2, 300 Ω; 5) ÉPP-09M2 potentiometer; 6) tungsten cathode; 7) cathode-unit screens; 8) sample of metal to be melted.

from the lower regions of the molten metal to the surface as a result of convective diffusion and the expulsion of the impurities as the single crystal becomes larger. In this way the sample is freed from impurities with melting points below that of the molybdenum. Hence the positive ion current arising in the course of zone melting is mainly related to the presence of impurities in the metal being refined.

Figure 1b shows the electrical circuit for measuring the ion current. The cap 1 of the melting equipment is insulated from the grounded body of the base by a rubber seal 2; this enables the cap to be used as a current collector for the ions. The inner surface of the cap is shielded by means of a wide-mesh grid 3 in order to suppress secondary-electron currents. The ion current flows between the collector and the grounded base by way of a potentiometer 4 with a resistance of 300 Ω, from which the voltage is taken to the recording device (ÉPP-09M2). The zone-melting ion-current characteristics were obtained with this apparatus.

The melting equipment is shown (Fig. 1b) in the form of a three-electrode vacuum device having a cathode 6, control electrodes 7, and an anode 8 constituting the metal subjected to remelting.

The ion-current characteristics representing the first and second meltings of a molybdenum sample containing various impurities are shown in Fig. 2a. The ion current (and hence the evolution of impurities from the zone of molten metal) bears a pulsed, nonuniform character. There are two typical forms of nonuniformities: current fluctuations with a frequency of 2.5 periods per second, and changes in the envelope with a period of about one minute.

The ion-current fluctuations with a frequency of 2.5 periods per second accompany the vigorous evolution of impurities from the zone during the first melting. The frequency of these fluctuations corresponds to periodic changes in the surface of the metal being melted, on which (after the first pass of the zone) transverse markings ("crests") are formed. The ion-current characteristic was subsequently obtained for the same samples on subjecting these to a second melting. The curve had a "quieter" appearance and indicated only slight separation of impurities.

The zone melting of refractory metals is often accompanied by a nonuniform structure of the resultant single crystals. These nonuniformities arise from the instability of the thermophysical parameters of electron melting; their traces have the form of thickenings of the single crystal, constrictions, and disruptions of the single-crystal structure. The changes in ion current reproduce the nonuniformity of the sample (Fig. 2b). The reason for this effect lies in the proportionality between the number of evaporating impurity atoms and the surface area of the molten metal. A reduction in the diameter (and hence a reduction in the area of the zone) leads to a sharp fall in ion current. This enables us to use the dependence of the ion

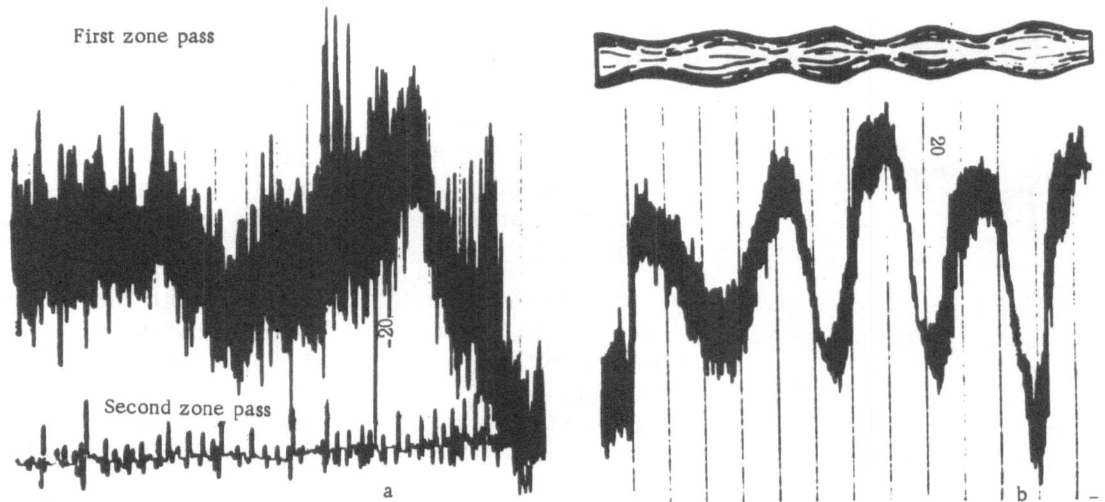

Fig. 2. Zone-melting ion-current characteristics for a uniform (a) and nonuniform (b) sample of molybdenum.

current on the diameter of the single crystal as an input parameter for a system of automatic control, regulating the diameter of the zone-melted sample.

Measurements show that the three-electrode system of the melting equipment has a static amplification factor of $\mu = 45$; it may accordingly be used as a basis for a servo system with negative feedback (relative to the ion current) in order to stabilize the diameter of the single crystal during growth.

The existence of a relationship between the ion current, the form of the ion characteristic, and the amount of impurity in the sample of metal being melted enables us to create a new method for controling (monitoring) the zone melting of molybdenum. The second parameter influencing the ion current (the area of the molten metal) under conditions in which a constant amount of power is fed into the molten zone enables us to set up a system of automatic control based on the diameter of the growing single crystal. In this way single crystals with a more regular linear structure may be obtained.

References

1. G. F. Zaboronok, G. I. Zelentsov, A. S. Ronzhin, and B. G. Sokolov, Electron Melting of Metals (1965).
2. É. Ya. Zandberg and N. I. Ionov, "Surface Ionization," Uspekhi Fiz. Nauk, 67:581 (1959).

STABILIZATION OF THE THERMOPHYSICAL CONDITIONS GOVERNING THE GROWTH OF SINGLE CRYSTALS OF REFRACTORY METALS

I. A. Brodskii, E. E. Petushkov, and S. M. Mikhailov

The mechanism of crystal growth lies in the tangential development of layers oriented along the planes having the greatest reticular density. This process is accompanied by the formation of a crystallization front, which largely determines the properties of the emerging single crystal.

The instability of the smooth surface of the crystallization front leads to the development of impurity–substructural boundaries, to an increase in the low-angle block disorientation, and hence to the appearance of a large number of dislocations and structural imperfections in the single crystal [1].

The instability of the smooth surface of the crystallization front depends on a number of factors, one of which is the stability of the thermophysical parameters of crystal growth in the zone-melting process. Satisfaction of the stability condition involves a number of difficulties, the chief of these being the unaccountable effect of the stream of ions arising from the zone of molten metal on the space charge of the cathode unit in the zone-melting equipment.

Attempts were made earlier [2, 3] to stabilize the conditions of zone melting by stabilizing the parameters of the electrical supply source. However, the systems then developed were insufficiently effective, in view of the fact that in various cases the melting conditions depended on changes in the internal resistance of the working gap in the melting equipment. In addition to this, the electron current in vacuo is not related by a directly proportional law to the voltage applied to the electrodes [4].

This leads to parametric oscillations of the anode current passing through the diode during the melting period, the parameter in question being the diameter of the zone of molten metal, while the period of the oscillations or fluctuations is governed by the thermal inertia of the heating system.

Thus stabilization of the parameters of the external electrical circuit is insufficient. In order to secure optimum conditions for the growth of single crystals, the parameters of the working gap in the melting equipment must also be stabilized.

Considering the zone-melting equipment merely in the form of an equivalent diode (Fig. 1a) fails to provide a complete picture of the causes of instability; these enter in a disguised form. The problem may best be treated by considering an equivalent circuit of the triode type (Fig. 1b).

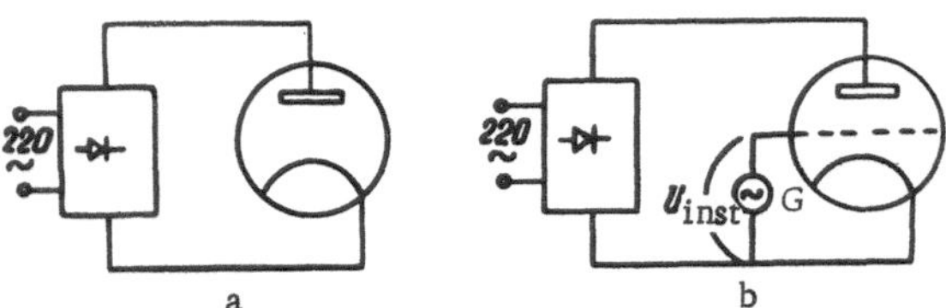

Fig. 1. Diode (a) and triode (b) equivalent circuits of the melting equipment.

The screens of the cathode unit (incorporated as a control electrode) are equivalent in their effect on the anode current to the control grid of a triode with an average amplification factor of $\mu = 45$, a mutual conductance of $S = 1.2$ mA/V, and an internal resistance of $R_i = 36$ kΩ.

Differences in the constructions of the cathode units result in no appreciable scatter of the parameters, the perveance remaining within the limits of $(1 \text{ to } 2) \cdot 10^{-6}\ \mathrm{A} \cdot \mathrm{V}^{-3/2}$.

The reason for the development of anode-current instability in this kind of triode lies in the changes taking place in the cathode potential minimum under the influence of the flow of ions from the molten anode. Thus the source of instability is, as it were, connected in series with the grid–cathode gap.

In the circuit of Fig. 1b, we may express the source of instability in the form of an equivalent generator G, incorporated in the grid circuit of the triode. By considering the circuit of the melting installation in this way, we reduce the instability of the anode current to the instability voltage U_{inst} of the equivalent generator, calculated from the formula

$$U_{inst} = \frac{\Delta I_{inst}}{S}$$

where ΔI is the change in the anode current,
S is the slope of the anode characteristic (mutual conductance).

The value of the instability, expressed in terms of the voltage of the equivalent generator, may be considered as a quantitative characteristic of the melting process as a whole, or as a partial characteristic representing the work of the cathode unit in the construction under consideration.

The following destabilizing (perturbing) factors influence the zone melting of metals:

1) Electrothermal feedback between the molten zone and the cathode through direct thermal radiation;

2) feedback between the metallic zone and the cathode through the ion current of the zone, which compensates the space charge of the cathode;

3) inconstancy of the cathode temperature;

4) changes in the properties of the cathode (contamination);

5) mechanical drift in the dimensions of the cathode unit, arising from the heating of the parts near the cathode and the zone of molten metal;

6) changes in the diameter of the sample.

Experience shows that the melting conditions are considerably affected by the operating mode of the cathode, which is determined by the relationship between the emission current and the cathode temperature (for a constant anode voltage), and is characterized by the three regions on the curve of Fig. 2a:

Region I — saturation of the anode current when the cathode is insufficiently heated; there is then no cathode space charge and the anode current is simply limited by the emissive capacity of the cathode.

Region II — nonequilibrium space charge with unstable anode current.

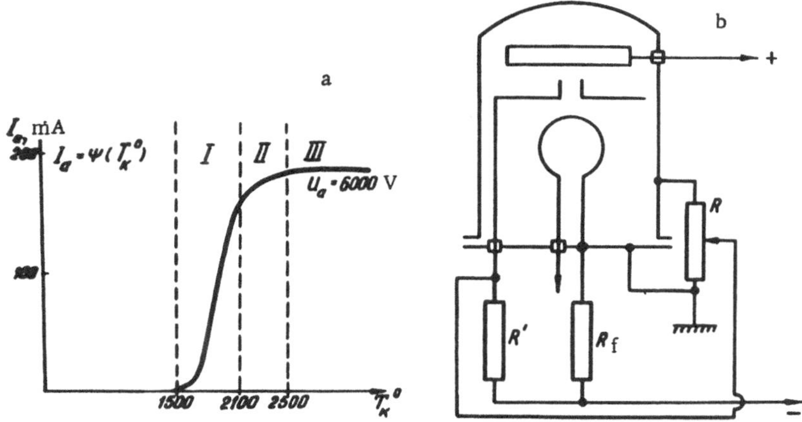

Fig. 2. Thermionic-emission characteristic of the cathode (a), and circuit for regulating the zone-melting equipment (b): R = 300 Ω (variable), SP-2; R_f (feedback) = 100 + 300 Ω, 60 W (noninductive); R' = 55 kΩ (noninductive).

Region III — equilibrium, substantial space charge; the anode current is now independent of the cathode temperature and is limited by the considerable potential dip in front of the cathode.

When the cathode operates in region III, the destabilizing action of the electrothermal feedback vanishes, but it becomes impossible to regulate the mode of melting by changing the cathode-heating current, as is usually done in the practical operation of existing zone-melting equipment. In place of this, it is sufficient to vary the anode voltage.

The destabilizing effect of ion-current feedback may be considerably weakened by using a cathode unit with spatial decoupling of the electron and ion currents (Fig. 3a). A characteristic of the construction of such a cathode unit is the formation of an open zone of molten metal, free from screens and traps, lying at a distance h below the cathode unit. This arrangement of the zone eliminates both the splashing of the cathode with heavy, drop-like surges of metal, and also the reflection of the same impurity-containing surges, these subsequently falling into the zone or the hot part of the sample. The reliable formation of an open zone is ensured by constructing the electrodes of the cathode unit on the principle of a right-angled triangle with opposite angles of 60 and 30° (Fig. 3a). The practical form of this (Fig. 3b) is a cylindrical shell 6, the body of the shell carrying the parts 8 needed for fixing the cathode unit to the guides of the zone-melting equipment. The shell 6 simultaneously constitutes a bearing structure for the upper and lower screens 2 and 4, and also the cylindrical surface required for shaping the electron beam. In the central part of the shell is the cathode 3, and in the side are the inspection windows 7.

This construction of the cathode unit ensures stability of the thermal conditions of the zone, prolonged service of the cathode, and a minimum impurity vapor pressure in the space close to the zone (i.e., no gas brake).

The use of such cathode units, together with stabilization of the anode supply voltage source, eliminates almost all the destabilizing factors and makes the zone melting of refractory metals semiautomatic, without manual adjustment of the conditions on the part of an operator.

As a result of the nonuniform distribution of impurities in the material being melted, and also random changes in the thermal conditions of melting, the growth of single crystals may take place in a nonuniform manner, with the formation of excrescences and constrictions. The

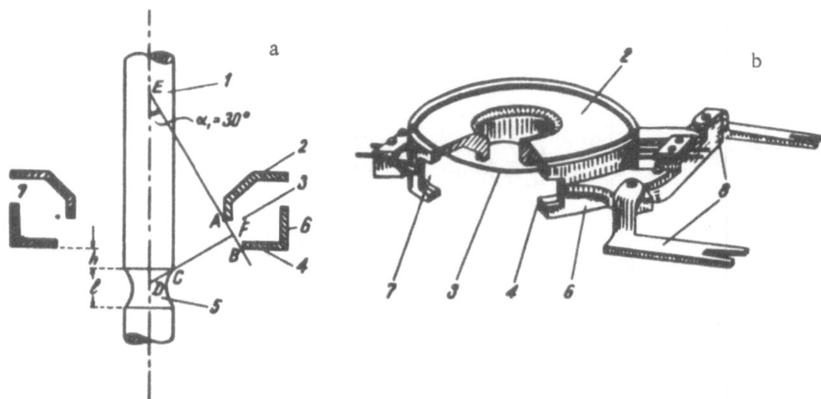

Fig. 3. Schematic representation of the focusing system (a) and structural realization of the cathode unit (b). 1) Sample of metal being melted; 2) upper focusing electrode and its principal point A; 3) tungsten cathode; 4) lower focusing electrode and its principal point B; 5) zone of molten metal, origin of zone at point C; 6) shell of cathode unit; 7) window; 8) fixing parts.

relation between the amount of impurities present and the ion current of the zone of molten metal enables us to construct an electrical automatic-control system (Fig. 2) for regulating the thermal conditions, such as to prevent the diameter of the growing crystal from changing.

By way of ion-current collector we use the cap of the melting equipment, which is insulated from the grounded base. The voltage drop in the resistance R is fed to the control electrodes of the cathode unit.

In order to reduce the control time constant [5], negative current feedback is introduced into the circuit by means of the resistance R_f.

The negative feedback introduces attenuation into the oscillatory processes arising from discharges between the electrodes of the melting equipment along the gas channels formed by the intensive ejection of impurities from the zone. The melting equipment incorporating negative feedback operates quietly, without electrical breakdowns or disruption of the mode of melting.

The measures taken to ensure internal stabilization of the parameters of the working gap in the melting equipment are only effective in combination with the stabilization of the anode-voltage supply source.

References

1. A. A. Kralina and V. O. Esin, Growth of Crystals, Vol. 5b, Consultants Bureau, New York (1967), p. 66.
2. Yu. M. Korobochkin, V. D. Pautov, and V. I. Shiryaev, Izv. Akad. Nauk SSSR, No. 1, p. 73 (1966).
3. O. V. Fedorov and L. S. Starostina, Pribory i Tekhn. Éksperim., No. 2, p. 156 (1963).
4. B. M. Tsarev, Calculation and Construction of Electron Tubes, GEP (1967).
5. L. I. Baida and A. A. Semenkevich, Electronic DC Amplifiers, GEP (1953).

ELEMENTS OF THE ENERGY CALCULATION OF HIGH-TEMPERATURE ZONE MELTING

A. A. Tserfas and E. E. Petushkov

In the process of zone melting the width of the molten section of the sample varies with the position of the heater; this influences the intensity of the purification and the stability of the molten section.

In an earlier paper [1] the width of the molten zone was calculated for various positions of the heater in the case in which heat transfer with the surrounding medium took place in accordance with Newton's law, the ends of the rod being thermally insulated. These calculations cannot be used for vacuum zone melting in the case of materials of a refractory nature, in view of the fact that thermal radiation and the outflow of heat through the ends of the rod are of major imporance in this case.

In this paper we shall derive a relationship between the width of the molten zone, the power supplied, the geometry of the system, and the thermophysical characteristics of the material in high-temperature melting.

Let us consider the energy balance of the molten zone. Figure 1 shows a section of vertical rod including a molten zone of height h between the phase boundaries S′ and S″.

Under steady-state conditions the energy balance for the molten zone is

$$W - Q_t' - Q_t'' = Q_u \tag{1}$$

in which W is the flow of energy conveyed to the region of the molten zone by the electron beam. In order to avoid unreasonable consumption of power, the electron beam is usually provided with narrow focusing; we may therefore consider that all the energy is fed into the region of the molten zone.

In Eq. (1) the quantities Q_t' and Q_t'' denote the flows of energy through the upper and lower interphase boundaries; Q_u is the flow of energy radiated from the surface of the molten zone in accordance with the Stefan-Boltzmann law. Using R to denote the radius of the rod, ε the emissivity, σ the emission constant, and T_{av} the average temperature of the melt, we obtain

$$h = \frac{W - (Q_t' + Q_t'')}{2\pi\varepsilon\sigma R T_{av}^4} \tag{2}$$

It should be noted that in this relation we have not considered the energy losses associated with the evaporation and ejection (splashing) of the molten metal; the temperature on the walls of the cooled chamber is considered low compared with the melting point of the metal.

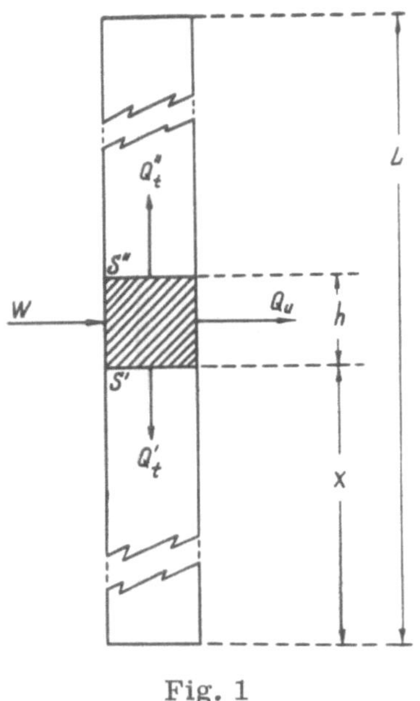

Fig. 1

The amount of energy passing from the molten zone along the solid part of the rod may, in accordance with the Fourier law, be expressed in the following manner:

$$Q_t = -\pi R^2 \lambda \left(\frac{\partial T}{\partial x}\right)_m \tag{3}$$

The direction of the coordinate axis is here taken as coinciding with the direction of heat flow. The temperature gradient in the solid parts of the rod near the upper and lower interphase boundaries may be determined by solving the differential equation of heat conduction, with heat transfer obeying the Stefan–Boltzmann law. According to earlier papers [2, 3] this equation has the form

$$\frac{d^2T}{dx^2} - \frac{2\varepsilon\delta}{\lambda R} T^4 = 0 \tag{4}$$

Using T_m to denote the melting point of the rod material in degrees Kelvin, T_K the temperature at the end of the rod, L the length of the rod, and X the distance from the lower end of the rod to the molten zone, we may write the boundary conditions as

$$\left.\begin{aligned} T(0) &= T_m \\ T(X) = T(L - X - h) &= T_K \end{aligned}\right\} \tag{5}$$

Making the substitution y = dT/dx we may use (4) to find

$$\frac{dT}{dx} = -\sqrt{\frac{4\varepsilon\sigma}{5\lambda R} T^5 + C_1} \tag{6}$$

where C_1 is a constant of integration.

Expressing (6) in the form $[T^5 + (C_1/\alpha^2)]\,dT = -\alpha\,dx$ and using only the first two terms of the rapidly-converging series into which the left-hand side of the equation may be expanded, after integration we have:

$$\frac{2}{3} T^{-3/2} \frac{C_1}{13\alpha^2} T^{-13/2} = \alpha x - C_2. \tag{7}$$

where C_2 is another constant of integration.

Using the boundary conditions we obtain

$$C_1 = 2\alpha^2 \frac{\alpha X + \beta}{\gamma} \tag{8}$$

where

$$\beta = \frac{2}{3}\left(T_m^{-3/2} - T_K^{-3/2}\right) \tag{9}$$

$$\gamma = \frac{2}{13}\left(T_m^{-13/2} - T_K^{-13/2}\right) \tag{10}$$

Substituting the value of the constant of integration from (2) into (6), we may express h in the form

$$h = p + \sqrt{p^2 - q} \tag{11}$$

in which

$$p = A - \frac{4}{25\,\alpha\gamma\,T_{\mathrm{av}}^8} \tag{12}$$

$$q = A^2 - \frac{4}{25\,\alpha^2\,T_{\mathrm{av}}^8}\left\{T_{\mathrm{m}}^5 + \frac{2}{\gamma}\left[\beta + \alpha(L - X)\right]\right\} \tag{13}$$

$$A = \frac{W}{2\pi\varepsilon\sigma R T_{\mathrm{av}}^4} - \frac{2}{5\,\alpha\,T_{\mathrm{av}}^8}\sqrt{T_{\mathrm{m}}^5 + \frac{2}{\gamma}(\beta + \alpha X)} \tag{14}$$

This result agrees with the empirical formula proposed by Belk [4] for metals with high melting points.

References

1. V. Ya. Frenkel' and B. A. Volchok, Inzh.-Fiz. Zh., 4:8 (1961).
2. Iain and Krishnan, Proc. Roy. Soc., A222:167-180 (1954).
3. H. Carslaw and D. Jager, Heat Conduction of Solids [Russian translation], Izd. Nauka, Moscow (1964).
4. Zone Melting, Izd. Nauch.-Issled. i Proekt. Inst. Prom. Redkikh Met., Moscow (1965).

ACOUSTIC ANISOTROPY OF METALLOCERAMIC (SINTERED METAL POWDER) MOLYBDENUM MOLDINGS

O. S. Kobyakov and E. E. Petushkov

When studying the velocity of longitudinal ultrasonic vibrations C_l in metalloceramic (sintered-metal) molybdenum moldings, we observed anomalous differences between the velocities of propagation in different directions (axial $C_{\parallel}$ and radial $C_{\perp}$). This difference in velocities may be regarded as characterizing an elastic acoustic anisotropy, which certain authors [1-3] associate with the molecular structure of the material, and, for certain rocks and coals in particular, with the existence of a layered structure. Acoustic anisotropy is usually expressed by the ratio of the velocities along the layers ($C_{\parallel}$) and across the layers ($C_{\perp}$):

$$A = \frac{C_{\parallel}}{C_{\perp}} \tag{1}$$

Frequently the ratio of the Young's modulus ($E_{\parallel}/E_{\perp}$) is used, this being equal to the square of the acoustic anisotropy [2]

$$K_a = \sqrt{\frac{E_{\parallel}}{E_{\perp}}} \tag{2}$$

The velocity of propagation of longitudinal waves (C_l) in an isotropic solid medium is determined by the elastic properties, the density, and the structure [4]

$$C_l = \left[\frac{(k + 4\mu)}{3\rho}\right]^{1/2} = \left[\frac{(1-\sigma)E}{(1-\sigma-2\sigma^2)\rho}\right]^{1/2} \tag{3}$$

where k is the modulus of bulk elasticity;
μ is the shear modulus;
σ is the Poisson's ratio;
ρ is the density;
E is Young's modulus.

The velocity of longitudinal waves (C_l) in samples in the shape of a cube with an edge of 28 mm was measured by a pulse method over a frequency range of 0.8 to 2.5 MHz. Samples previously subjected to pressing, sintering, welding, and forging were studied. The measurements showed (Fig. 1) an almost complete absence of anisotropy in the pressings ($C_{\parallel}/C_{\perp} = 1$), the average velocity in the pressings being $C_{\parallel} = C_{\perp} = 1500$ to 1700 m/sec for a mean density of 4.4 to 4.7 g/cm^3. Sintering has no effect on the anisotropy of the properties ($C_{\parallel}/C_{\perp} = 1$), al-

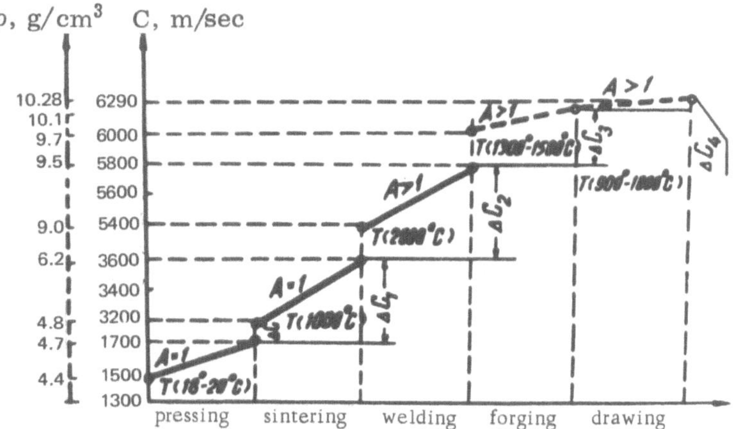

Fig. 1. Diagram representing the growth in mean density and velocity ($C_{\parallel}$ and $C_{\perp}$) of longitudinal (C_l) vibrations as a result of the preliminary treatment of molybdenum.

though there is a jump in velocity ΔC_1 ($C_{\parallel} = C_{\perp} = 3200$ to 3600 m/sec) with a simultaneous rise in density to between 4.8 and 6.2 g/cm³.

As a result of welding the velocity increases (ΔC_2) again and reaches between 5400 and 5800 m/sec in the axial direction and between 4650 and 5000 m/sec in the radial direction, with a mean density of 9 to 9.5 g/cm³. After welding the ratio $C_{\parallel}/C_{\perp}$ is greater than unity ($A > 1$), i.e., the metal acquires anisotropic properties. Measurement of forged samples reveals a further increase in velocity (ΔC_3) to $C_1 = 6000$ to 6150 m/sec, again with a distinct anisotropy ($A > 1$) and a mean density of 9.7 to 10.12 g/cm³. After drawing, the anisotropy coefficient remains at $A > 1$, while the velocity increases so as to approach the maximum of

$$C_l = 6290 \text{ m/sec}$$

In the case of isotropic samples (those subjected to pressing and sintering), the velocity variation in the moldings is represented by a velocity circle (Fig. 2a, curves 1, 2), while in the case of anisotropic samples (welded material) it is represented by an ellipse (Fig. 2a, curve 3)

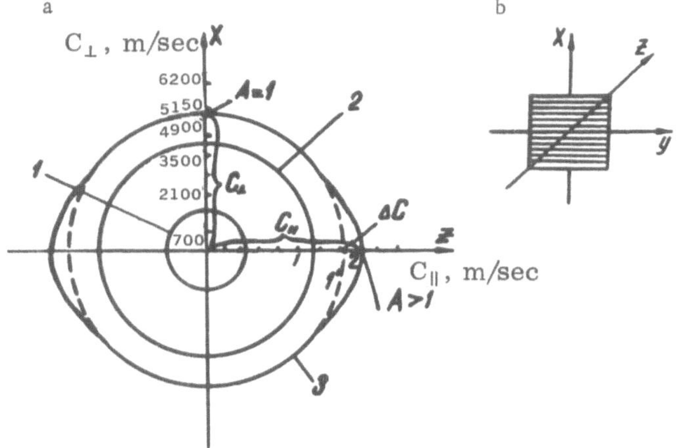

Fig. 2. Velocity circles and ellipses (a) for samples sounded along the x and z axes (b); x = radial sounding axis; y = axial sounding.

with the velocity C_l predominating along one of the sounding axes (z). The welding operation has a considerable effect on the relative shrinkage characteristics; the shrinkage in the axial direction is 7.5 to 8% and in the radial direction 10 to 12%, the mean relative density in individual parts of the samples (end to middle, surface to center) varying from 14.7 to 7%.

As a result of a large number of measurements (over 100), we obtained the corresponding values for the acoustic anisotropy coefficient. For welded metalloceramic molybdenum subjected to longitudinal vibrations (C_l) the coefficient lay between 1.01 and 1.3.

The conditions of heat treatment may be directly related to the anisotropy coefficient. On increasing the welding current from 4700 to 5000 A the anisotropy coefficient A increases from 1.011 to 1.036.

References

1. N. I. Brazhnikov, Ultrasonic Methods, Izd. Énergiya, Moscow (1965), p. 29.
2. E. G. Martynov and A. K. Matveev, Use of Ultrasonics in the Study of Materials, No. 10, Moscow (1960), p. 135.
3. M. P. Volorovich and E. Bayuk, Use of Ultrasonics in the Study of Materials, Vol. 11, Moscow (1960), p. 147.
4. B. Carlin, Ultrasound [Russian translation], IL, Moscow (1950), p. 259.

DETERMINATION OF HYDROGEN IN MOLYBDENUM BY A DIFFUSION-MANOMETRIC METHOD

E. E. Petushkov, A. A. Tserfas, and T. M. Maksumov

The most convenient method of determining traces of hydrogen in metals is the method of vacuum heating. The metal sample is placed in a quartz exchange-reactor and the air is pumped out of this. For temperatures of up to 650°C, only hydrogen evolves from the metal [1]; for higher temperatures, up to 1100°C, carbon monoxide, nitrogen, and carbon dioxide also appear [2]. For example, mass-spectrometric analysis of the composition of the gas evolved from a sample at t = 900°C gave the results: H_2 — 70.3%; CO — 24.3%; CO_2 — 0.27%; and nitrogen 5.03%.

The rate of evolution of hydrogen from the sample depends largely on the temperature; thus in order to ensure complete elimination of the gas the metal samples are held at a temperature of the order of 1000°C. However, in this case no simple manometric determination of the amount of hydrogen present can be carried out, as the other gases present may contribute over 40% to the total gas content.

The dependence of the hydrogen content of molybdenum on the gas pressure in the surrounding medium may be expressed in the following manner, as in the case of other group A metals:

$$S = S_0 \sqrt{P_{H_2}} \tag{1}$$

where S_0 is the solubility of hydrogen in molybdenum at atmospheric pressure in $cm^3 \cdot atm/100\,g$; P_{H_2} is the pressure of hydrogen in the surrounding medium, in atm.

According to earlier data [3], at 1000°C and a pressure of 10 mm Hg the equilibrium amount of hydrogen in the metal equals $5 \cdot 10^{-6}$%; the pressure may be reduced as a result of an increase in the volume of the reaction chamber.

In order to analyze the gas mixture emerging from the sample, possible methods include those based on chromatography, polarography, and mass-spectrometry; however, these are too complex to be very convenient.

In order to determine the amount of hydrogen evolved from the sample we accordingly took advantage of the diffusion characteristics of the material. For this purpose we employed a leak system comprised of a nickel tube sealed at one end with a wall thickness of 0.1 mm. The tube was mounted in a glass cylinder and the test gas was passed into the latter. The nickel tube was heated to the required temperature by passing an electric current through corresponding electrodes, the temperature being monitored with a thermocouple. The gas mixture under

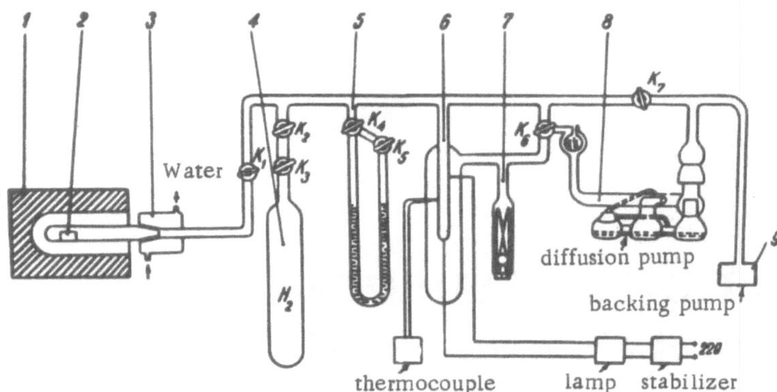

Fig. 1. Arrangement of the apparatus for determining hydrogen impurity: 1) muffle furnace; 2) metal sample in quartz exchanger; 3) water jacket; 4) gas cylinder; 5) manometer; 6) leak; 7) thermocouple tube; 8) diffusion pump; 9) backing pump.

consideration was passed into the leak device which had first been evacuated and heated. Since the diffusion coefficient of hydrogen was much greater than that of the other gases it could safely be assumed that only the hydrogen passed through the nickel.

The general arrangement of the apparatus for determining traces of hydrogen appears in Fig. 1. The exchanger is a quartz tube 25 to 30 mm in diameter and sealed at one end, connected by a ground-glass joint to the pumping system; the joint has a running-water cooling jacket. In the glass cylinder 4 is a store of pure hydrogen, used for calibration. The oil manometer 5 is used to measure the total pressure of the gases evolving from the sample and also the pressure of the pure hydrogen during calibration. The part of the volume comprising the inside of the exchanger and the sections of leak and tubes bounded by the taps K_2, K_4, K_6, we may arbitrarily call system A. The region of space incorporating the monometric tube LT-2, the air conduit bounded by the tap K_6, and the inside of the leak on the side to which the hydrogen is diffusing constitutes system B.

The system is evacuated with an oil diffusion pump. By using K_7, system A may be evacuated with the backing pump alone. In order to heat the quartz exchanger a muffle furnace heated to 1000°C is placed around this. The apparatus is evacuated, tested for absence of leaks, and calibration commences. After establishing a pressure of 10^{-4} mm Hg in systems A and B and heating the nickel filter to the specified temperature, pure hydrogen is admitted from the gas cylinder 4 by means of the taps K_2 and K_3, which in concert with the tube joining them constitute a simple dosing system. In this way the initial pressure P in the system A is established and a stop-watch is used in order to determine the time $\Delta\tau$ required for the pressure in the system B to rise from P_1 and P_2. The time required to create the pressure P_1 is chosen in such a way as to ensure that the front of the gas diffusing through the barrier should be able to penetrate through the whole thickness; this eliminates the effect of the initial hydrogen distribution within the barrier on the results of the experiment. The pressure P_2 is set arbitrarily.

According to the earlier treatment [4], the average time for the diffusing atoms to penetrate through a distance h in the metal is given by the formula

$$\overline{\Delta\tau} = \frac{h^2}{2D} \tag{2}$$

where h is the depth of penetration, in cm;
D is the diffusion coefficient, in cm^2/sec.

These measurements are employed in order to plot a calibration graph of $\Delta\tau$ against P for use in determining the partial hydrogen pressure on admitting a gas mixture into system A.

In determining the hydrogen content of a molybdenum sample, the latter (sample weight G) is first placed in the quartz exchanger. After evacuating to a pressure of about 10^{-4} mm Hg with the tap K_1 closed, the exchanger is held at 1000°C for a period sufficient to establish an equilibrium state, this period being determined experimentally, and constituting approximately 40 min for molybdenum samples 6 mm in diameter.

After completing the heat treatment and cooling the exchanger, the tap K_1 is opened and the partial pressure of the hydrogen in the volume V_a is measured. The amount of hydrogen evolved equals

$$G_{H_2} = \frac{\gamma_{H_2} \cdot \left(V_A - \frac{G}{\gamma_{Mo}} \right) \cdot (P - P_x)}{760} \quad (3)$$

where γ_{H_2}, γ_{Mo} are the specific gravities of hydrogen and the molybdenum samples, in g/cm^3;
P_x is the pressure of the hydrogen evolved from the walls of the quartz exchanger, in mm Hg.

In the actual experiments P_x proved to be smaller than the level of experimental error, thus justifying its omission.

The sensitivity of the measurements is mainly determined by the sensitivity of the procedure employed for measuring the rate of hydrogen diffusion through the nickel filter. The use of a manometer tube of the LT-2 type together with a VIT-1 apparatus gives a sensitivity of the order of $5 \cdot 10^{-5}$% for a sample weight of 30 to 40 g. The relative error in the measurements for a hydrogen content of $5 \cdot 10^{-4}$% is 5%.

References

1. Yu. A. Karpov and G. G. Glavin, Analysis of gases in metals, Zavod. Lab., 31:2 (1965).
2. Yu. A. Klyachko, L. L. Kunin, S. P. Fedorov, and I. N. Larionov, Interaction of gases with metals, in: Analysis of Gases in Metals, Izd. AN SSSR, Moscow (1960).
3. S. Dushman, Foundations of Vacuum Technology [Russian translation], IL, Moscow (1950).
4. W. Zeit, Diffusion in Metals [Russian translation], IL, Moscow (1958).
5. Z. M. Turovtseva and L. L. Kunin, Analysis of Gases in Metals, Izd. AN SSSR, Moscow–Leningrad (1959).

MOSAIC STRUCTURE OF MOLYBDENUM SINGLE CRYSTALS

R. N. Alimova and G. A. Klein

One of the most efficient methods of growing oriented single crystals from refractory metals, which has become very frequently employed in the last few years, is the method of vacuum electron-beam zone-melting. It is naturally interesting to secure data relating to the mosaic structure of such refractory-metal single crystals and to determine the influence of various factors on the degree of perfection of the crystals in question.

In this paper we shall present the results of a recent investigation into the mosaic structure of molybdenum single crystals. The samples for study constituted molybdenum single crystals grown by the method of electron-beam zone melting (using a seed oriented in the [100] direction) from metalloceramic molybdenum moldings, involving two zone passes, the first at a velocity of 4 mm/min in order to provide zone refining, and the second at 2 mm/min in order to grow oriented single crystals on the seed. These single crystals constituted cylindrical bars 14 to 18 mm in diameter and 320 to 350 mm long. For preparing the samples we used four single crystals oriented in the [100] direction, the axes of these deviating from the specified crystallographic direction by no more than 3 to 5°.

The single crystals were cut into cylinders 25 mm long in a direction perpendicular to the axis. For purposes of examination we used a shear surface, first ground with a succession of emery papers and then electropolished for 20 sec in a mixture of 90% sulfuric acid and 10% methyl alcohol at a voltage of 20 to 30 V and a current density of 1.6 A/cm^2. In all cases the orientation of the single crystals and the microsection samples prepared from these was determined by the Laue back-reflection x-ray method.

In order to reveal the microstructure in the MIM-8 metallographical microscope, the microsections so prepared were subjected to chemical etching for 2 to 3 sec, using 30 g of red salt and 10 g of caustic soda in 100 cm^3 of distilled water.

For a preliminary qualitative determination of the degree of perfection (mosaic structure) of the single crystals, we used the Laue-spot x-ray diffraction topography method. In Laue back-reflection photography the reflection angles are close to 90°, i.e., $\sin \Theta \sim 1$, Hence the Laue-spot method distortionlessly reproduces a plane section of the pencil of normals. The shape of the spot depends on the tube focus, and the internal structure on the perfection of the single crystal under examination. If the crystal is quite perfect, the spot is strictly localized and uniformly darkened, its size being determined by the conditions under which the photograph is taken; if the crystal is made up of disoriented fragments (macroscopic blocks), the spot also consists of several spots, the total size of the Laue reflection being determined by the maxi-

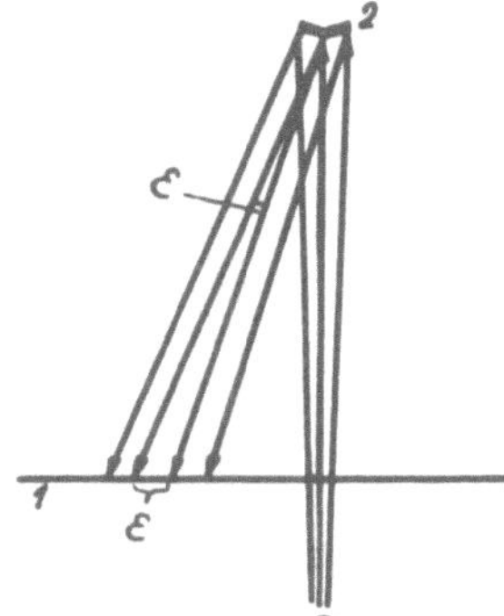

Fig. 1. Geometry for taking Laue topograms using a wide x-ray beam: R = x-ray tube; 1) film; 2) sample.

mum block disorientation. We obtained our Laue back-reflection pictures with a KROS1 camera, using 1 and 0.6 mm diaphragms, in molybdenum radiation, the x-ray tube current and voltage being 10 mA and 45 kV respectively.

This method of determining the perfection of single crystals by reference to the shape of the Laue spots is insufficiently objective, owing to the limited nature of the samples studied. In order to obtain fuller information as to the internal mosaic structure of the single crystals, we used a modified method involving an examination of the topograms of Laue reflections obtained with a widely-diverging beam of polychromatic x radiation from a standard BSV-2 x-ray tube with a molybdenum anode [1]. In this case also the photographs were taken with a KROS1 camera, the distance between the sample and film being increased to 100 mm and that between the x-ray tube focus and sample to 130 mm. The sample surface was irradiated with a wide, divergent beam, so that the irradiated part of the surface equalled approximately 16 mm^2 for a voltage of 20 kV and a current of 20 mA. When two subsidiary grains disoriented by an angle ε fell into the region of irradiation, the reflections from these were also disoriented by an angle ε (Fig. 1).

By considering the x-ray diffraction topograms obtained by means of the wide beam in back-reflection geometry (Fig. 2), we may quantitatively estimate the disorientation angles be-

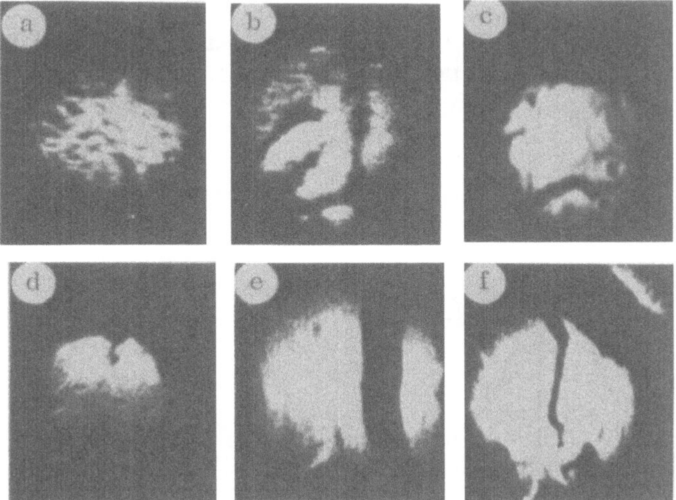

Fig. 2. Topograms of Laue (100) reflections from the surface of a microsection obtained from a molybdenum single crystal, using a wide x-ray beam. The individual pictures were obtained from different points.

tween the macroscopic blocks from the formula $\varepsilon = n/R$, where ε is the disorientation angle, n is the width of the line on the topogram, and R is the distance from the sample to the film. The smallest angular resolution under these conditions of photography is approximately 0°3′.

In order to determine the average mosaic angle we used the double-crystal-spectrometer x-ray method, the recordings being taken with a standard URS-501 diffractometer. According to an earlier investigation [3] the mosaic angle may be most simply determined by rotating the crystal, leaving the counter slit width constant and having a narrow source slit. The primary beam was monochromatized by reflection from a quartz crystal monochromator; it passed through a slit 0.25 mm wide and fell into a 4-mm wide Geiger counter. The integral-intensity curves were plotted from the counter readings by rotating the sample in steps of 0°1′ over a range of 1° around the Bragg reflection angle for the (100) plane.

A study of the internal macroscopic substructure of molybdenum single crystals based on the examination of microsections in the metallographical microscope showed that substructures of three orders usually occurred. The surface of a microsection constituting a cross section of the single crystal clearly revealed sharply delimited first-order subsidiary grains, drawn out along a particular direction. In the majority of cases this direction coincided with one of the [110] crystallographic directions. The large blocks were in turn divided into finer subsidiary grains by a rather fainter network of boundaries (second-order blocks). Inside the larger second-order blocks was a further cellular structure comprising blocks of the third order.

The macroscopic substructures were studied by the ordinary Laue back-reflection x-ray method in selected parts of the cross section of the molybdenum single crystal, a practically perfect germanium single crystal being used as control sample.

The Laue spots obtained from the control sample were strictly localized and uniformly darkened. If the irradiated part of the molybdenum single crystal contained subsidiary boundaries of the first order, the Laue spot was split sharply in two; if the Laue reflection were obtained from a region containing subsidiary boundaries of the second order, a spot with light and dark regions in it appeared on the topogram, indicating the presence of a number of small, slightly disoriented blocks at this point.

By using a wide beam for the Laue back-reflection photography, we obtained sharper topograms (Fig. 2), clearly revealing the strips corresponding to subsidiary boundaries of the first and second orders. When several subsidiary boundaries fall into the region of irradiation, the topogram shows a series of small spots (Fig. 2a). The observed disorientation between the

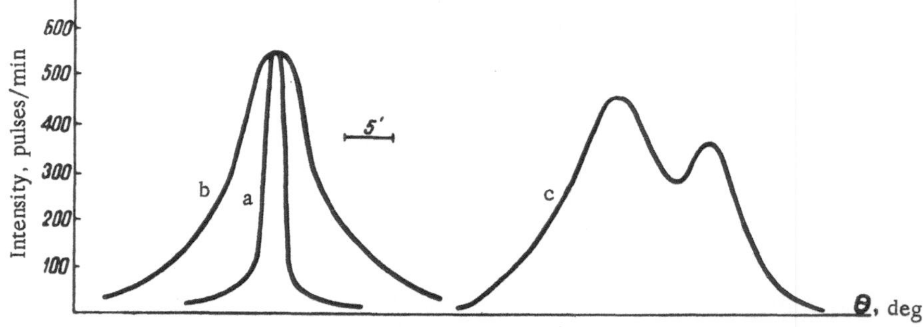

Fig. 3. Integral-intensity curves obtained by the method of the double-crystal x-ray spectrometer: a) standard calcite crystal; b, c) (100) surface of a microsection obtained from a molybdenum single crystal.

subsidiary boundaries lies between 0° 5′ and 1°. For example, the band in Fig. 2e corresponds to a disorientation of 0° 50′. Figure 2d shows two sharp bands, one wide and the other narrow, corresponding to disorientations of 1° and 0° 10′ respectively. In Fig. 2a the spot is broken up into several spots by a series of subsidiary boundaries with a disorientation averaging between 0° 5′ and 0° 10′.

The results confirm those obtained for the same samples of molybdenum single crystals with the double-crystal x-ray spectrometer. The disorientation of the second-order blocks found in this way, corresponding to the half-width of the integral-intensity curves, was 0° 10′ to 0° 15′. Curve a in Fig. 3, representing the integral intensity of a standard crystal, had a half width of 3′. If a first-order subsidiary boundary fell into the region examined, the intensity curve consisted of two peaks with a disorientation between them. Thus Fig. 3 illustrates one particular integral curve c, consisting of two peaks with a disorientation of 0° 20′ between them and a total half-width of about 0° 50′.

References

1. V. O. Esin and T. V. Ushkova, in: Growth and Imperfection of Metallic Crystals, Izd. Naukova Dumka, Kiev (1966), pp. 278-289.
2. D. M. Kheiker and L. S. Zevin, X-Ray Diffractometry, Fizmatgiz, Moscow (1963), p. 354.

CHANGES IN THE STRUCTURE OF MOLYBDENUM SINGLE CRYSTALS SUBJECTED TO THERMAL FATIGUE

V. A. Krakhmalev and G. A. Klein

Refractory metals such as Ta, Mo, W, Re, Nb, and others are attracting more and more attention in various fields of science and technology as a result of their heat resistance and strength at high temperatures. The physicomechanical [1, 2], chemical [3-5], and emission [6, 7] properties of these materials are being extensively studied. However, the thermal fatigue of refractory metals, i.e., the changes taking place in the physicomechanical and other characteristics on subjection to periodic thermal cycling have not so far been adequately considered. This applies particularly to single crystals obtained from refractory metals.

The most abundant data available relates to changes taking place in the shape of polycrystalline metals as a result of their periodic deformation (slip, twinning, etc.) and the swelling of the materials associated with thermal cycling [8]. However, the behavior of the dislocations in the course of thermal fatigue, particularly in the early stages, has been insufficiently studied, although it is well known that almost all the strength characteristics depend to a certain extent on the dislocation distribution established.

It is not always possible to determine the changes taking place in the dislocation structure of metal samples during thermal fatigue, for example, by means of dislocation etching. The difficulties encountered in revealing dislocations in this way arise from the limited number of etchants available, many of which in fact only reveal dislocations in certain metals and by no means in all crystallographic planes [9]. Furthermore, sometimes even an established etchant fails to reveal changes taking place in the dislocation structure as a result of thermal cycling, such etchants being insensitive to "fresh" dislocations.

In this respect molybdenum is an exception, as there are some well-known etchants available which will reveal dislocations even under conditions of thermal cycling. This enables us to follow the dynamics of the dislocation changes in the very early stages of thermal fatigue (almost from the first to the last thermal cycle) and establish corresponding regular laws.

In order to exclude the influence of mosaic substructure, molybdenum single-crystal samples 30 to 35 mm and 10 to 12 mm long were cut from the same single-crystal bar, 14 mm in diameter, obtained by electron-beam zone melting, and subjected to thermal cycling in two modes: the first involved heating to 700°C and cooling to 100°C in a vacuum of 10^{-5} mm Hg, and the second involved heating to 700°C in a muffle furnace in air and cooling in water to between 10 and 15°C.

The temperature was measured with thermocouples. No special measures were taken to prevent oxidation. The molybdenum single crystals 10 to 12 mm long were cycled in the first

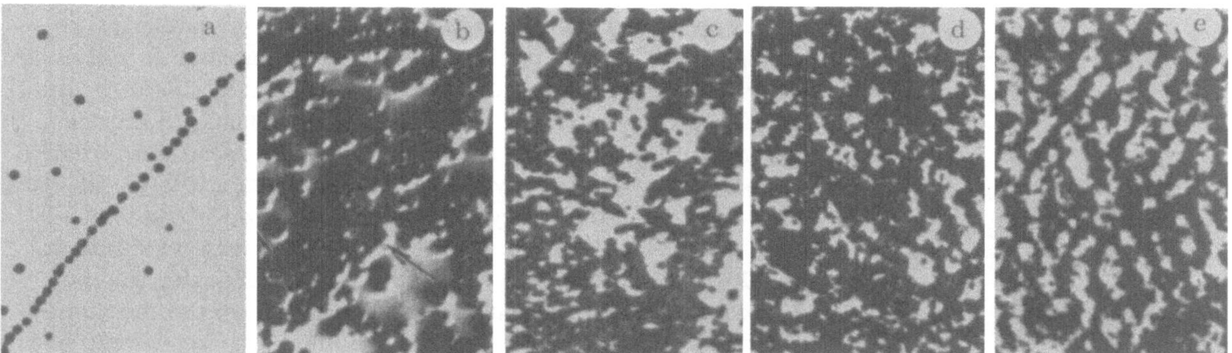

Fig. 1. Dislocation structure of molybdenum single crystals: a) control samples (magnification 510); b) after 150 thermal cycles (510); c) partial polygonization of the single crystals after 60 thermal cycles (300); d) after 500 thermal cycles (510); e) after 1540 thermal cycles (510).

mode and the longer (30 to 35 mm) ones in the second. The heating time in the first mode was 2 min and the cooling time 19 min. The heating and cooling times in the second mode were 105 and 15 sec respectively.

In the molybdenum single crystals cycled in the first mode the dislocation structure was studied after each cycle up to 10 and then after every ten cycles. The samples cycled in the second mode were studied by dislocation etching and x-ray analysis, microhardness measurements and electron-microscope examination (replica method), and also by microfractography, after 500, 1540, 3500, 7000, and 10,000 thermal cycles. The microhardness measurements, the x-ray diffraction pictures, and the replica photographs all related to the polished {100} planes of the molybdenum samples. The dislocation etching of the {100} planes was carried out in concentrated sulfuric acid with a small quantity of methyl alcohol. The microhardness was measured on the PMT-3 hardness tester; the "epigrams" (x-ray photographs) were taken in the URS-55 instrument, using the Laue technique.

The results show that very substantial changes in crystal structure take place during the thermal fatigue of molybdenum single crystals, sharply affecting the physicomechanical properties of the material. While remaining single-crystal, the molybdenum samples were completely polygonized after thermal cycling. The polygonization process took place in the following manner.

The control samples of molybdenum single crystals (after the etching of their polished surfaces) had a typical mosaic structure consisting of a series of blocks. The mosaic blocks had very different dimensions. Some of the mosaic blocks could only be seen under the microscope, others with the unaided eye. The mosaic blocks of the control molybdenum samples had a small mutual disorientation, since their boundaries appeared in the form of chains of etch pits in the course of dislocation etching (Fig. 1a). Inside the blocks the dislocation distribution was quite random.

From the very first cycles of thermal action, the dislocation density in the molybdenum single crystals increased sharply. As a result of the multiplication and movement of the dislocations, the subsidiary block boundaries started becoming diffuse (Fig. 1b). As the number of thermal cycles increased, the dislocation density in the samples became so great that after etching, the {100} planes of the molybdenum appeared completely black, owing to the large number of merging etch pits. The old subsidiary block boundaries vanished, gradually losing their initial individuality characteristic of the period before thermal cycling had started (Fig. 1a).

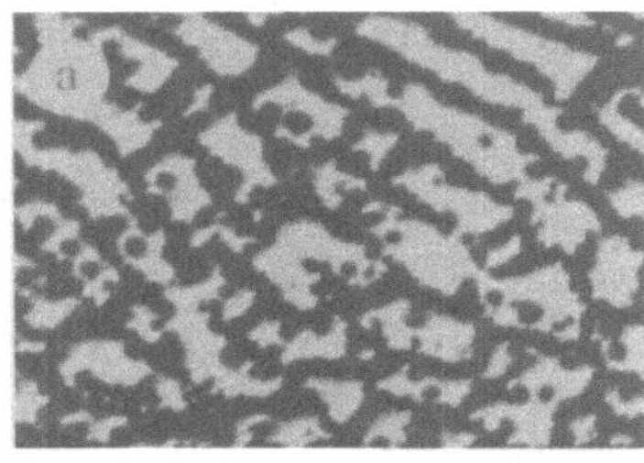

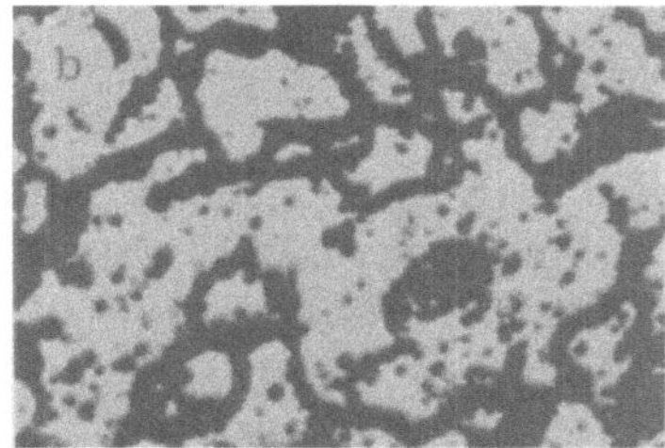

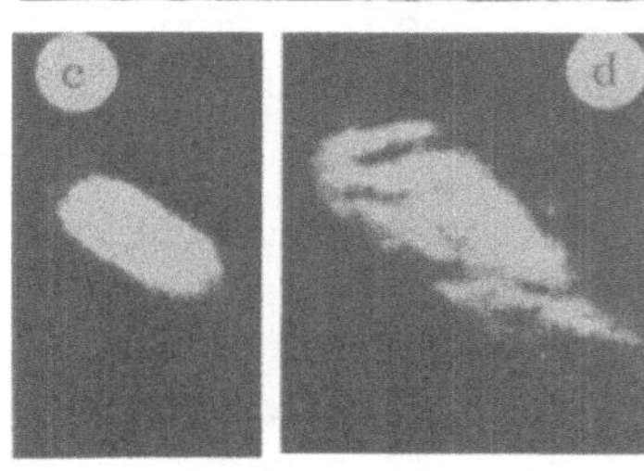

Fig. 2. Dislocation structure of molybdenum single crystals after 3500 (a) and 10,000 (b) thermal cycles, and structure of the x-ray pictures obtained from the control samples (c) and from molybdenum samples after experiencing 70,000 thermal cycles (d). Magnification in a and b: 510.

The appearance of the first signs of polygonization resulting from thermal cycling (and also the manner in which the latter took place) depended greatly on the rate of cooling the samples in the course of thermal cycling. Thus, in the molybdenum single-crystal samples subjected to the first mode of treatment, partial polygonization occurred at the 60-th cycle (Fig. 1c). The single-crystal samples treated in the second mode were completely polygonized over the whole volume at the 500-th thermal cycle (Fig. 1d). With increasing number of thermal cycles, the partial polygonization of the molybdenum single crystals increased in volume until the sample had polygonized completely.

We see from Fig. 1d that, after 500 thermal cycles, the polygons were surrounded with dislocation boundaries of considerable thickness. After 1540 cycles (Fig. 1e) the dimensions of the polygons increased, while the walls of dislocations diminished. The polygons became more or less of regular form. In the epigrams (x-ray photographs) taken from these samples the reflections had a distinct asterism; the x-ray diffraction spots became banded. The discontinuous structure of the spots indicated that the incident x-ray beam had embraced a certain number of blocks of the single crystal, with a discretely varying orientation.

After 3500 thermal cycles of the molybdenum single crystals, etching revealed very distinct polygonization (Fig. 2a). The polygons had the form of rectangles of various sizes, the boundaries consisting of chains of etch pits. As the number of thermal cycles increased, the boundaries of the polygons in the single crystals became more and more disoriented and curved, while the polygons increased in size (Fig. 2b). In the internal structure of the x-ray reflections there was an increase in the intensity and the gaps between the bands comprising the x-ray diffraction spots (Fig. 2c, d).

Whereas the x-ray spots of the control samples (Fig. 2c) had a regular form and uniform darkening, those of the molybdenum samples subjected to thermal cycling exhibited a banded structure (Fig. 2d). Thus the asterism of the x-ray spots obtained from the same thermally cycled samples of molybdenum single crystals as those in which the dislocation structure was analyzed also confirms the occurrence of polygonization. As a result of this, for example, the hardness of the single crystals after 500 thermal cycles increased by 25.6%.

The polygonization of molybdenum single crystals after slight deformation and high-temperature annealing at 2200 to 2300°C was also studied by L. N. Larikov, O. E. Zasimchuk, and Zh. Ya. Kutikhina [10], using x-ray and microstructural methods.

Calculation of the angles between the boundaries of the polygons and the traces of the cleavage planes {100} and {110} showed that the dislocation boundaries of the polygonal lattice lay in the molybdenum slip planes {110}, {112} and {123}.

An electron-microscope study of replicas separated from the polished {100} surfaces of thermally cycled molybdenum single crystals revealed the presence of second-phase precipitates. The impurity was precipitated along the dislocation lines. The impurity precipitates (apparently molybdenum carbides) lay in the {100} planes and reproduced the features of the polygon boundaries. We may suppose that the copious precipitation of impurity particles in the matrix of the molybdenum single crystals was caused by considerable contamination of these during the electron-beam crystal-growth process, which was carried out in a vacuum produced by means of oil-vapor pumps without using oil-vapor "traps."

References

1. N. V. Ageev, A. A. Babarěko, G. E. Chuprikov, and N. N. Bokareva, Izv. Akad. Nauk SSSR, Metally, No. 4, pp. 84-89 (1966).
2. P. Beardmore and D. Hull, J. Less-Common Metals, 9(3):168-180 (1965).
3. R. E. Pawel, S. V. Cathcart, and S. S. Campbell, Acta Metallurg., 10(2):149-160 (1962).
4. R. W. Bartlett, J. Electrochem. Soc., 112(7):744-746 (1965).
5. R. W. Bartlett and J. W. McCamont, J. Electrochem. Soc., Vol. 112, No. 2, Pt. 1, 148-152 (1965).
6. G. N. Shuppe, Electron Emission of Metal Crystals, Izd. SAGU, Tashkent (1959).
7. U. A. Arifov, Interaction of Atomic Particles with Solid Surfaces, Izd. Nauka, Moscow (1968).
8. N. N. Davidenkov and V. A. Likhachev, Irreversible Deformation of Metals Under Thermal Cycling, Mashgiz, Moscow–Leningrad (1962).
9. V. S. Ivanova, Fatigue Failure of Metals, Metallurgizdat, Moscow (1963).
10. L. N. Larikov, O. E. Zasimchuk, and Zh. Ya. Kutikhina, Ukrain. Fiz. Zh., 10(8):899-904 (1965).

EFFECT OF NEUTRON IRRADIATION ON THE STRUCTURE AND MICROHARDNESS OF DIFFERENT FACES OF MOLYBDENUM SINGLE CRYSTALS

G. N. Grishkov, D. I. Mavridi, and G. A. Klein

Irradiation with neutrons of fairly high energy leads to the formation of lattice defects in the crystal structure of metals and other solid materials and thus changes the physical and mechanical properties of the latter. Little work has been done in this connection with metallic single crystals, particularly those of the refractory metals [1].

We studied the effect of neutron irradiation on the structure and physicomechanical properties of different faces of molybdenum single crystals, this being one of the most promising refractory metals in the new technology. Molybdenum single crystals were grown from metalloceramic (sintered-metal) bars by electron-beam zone melting in vacuo, using seeds oriented in a specific crystallographic direction. The desired orientations of the planes were achieved by reference to Laue photographs taken by the back-reflection method. Plates 1.5 to 2.0 mm thick and 18 to 22 mm in diameter were prepared with the orientation of the (100), (110), (111), and (112) planes, the surfaces being carefully polished. The microhardness was determined on a PMT-3 hardness tester. Some of the samples were vacuum-annealed at $2 \cdot 10^{-5}$ mm Hg and 1000°C for 5 h, after which the microhardness fell by 11% for the (100), 6% for the (112), and 5% for the (110) plane.

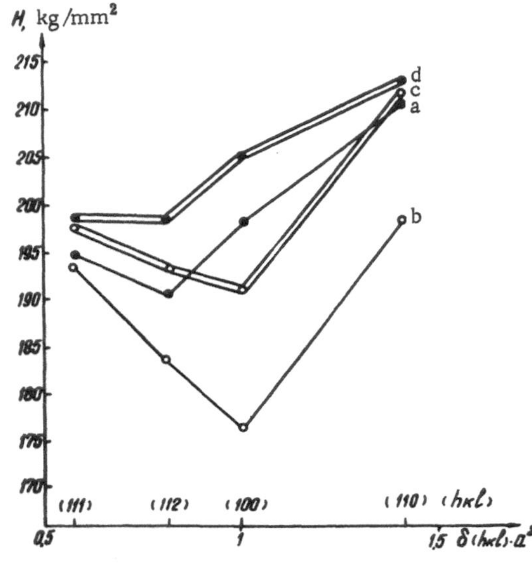

Fig. 1. Microhardness of Mo single crystals as a function of the crystallographic indices (hkl) and the packing density of the atoms on the face ($\delta_{(hkl)} a^2$) [3]: a) unannealed and nonirradiated samples (control); b) annealed samples (T ≡ 1000°C, t = 5h, P = $2 \cdot 10^{-5}$ mm Hg); c) irradiated samples ($2 \cdot 10^{18}$ neutrons/cm²) after annealing; d) similarly irradiated but unannealed samples.

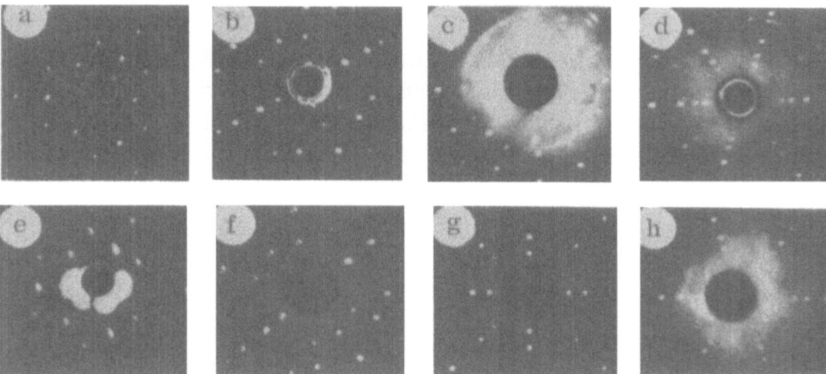

Fig. 2. Laue photographs taken from the faces of molybdenum single crystals before and after neutron irradiation with a dose of $2 \cdot 10^{18}$ neutrons/cm^2. Nonirradiated (control) samples: a) (111); b) (112); c) (100); d) (100); irradiated samples: e) (111); f) (112); g) (100); h) (110).

In the case of the (111) plane the microhardness hardly changed at all, possibly as a result of the lower mobility of the defects on this face, these defects being poorly annealed. It was shown earlier [2] that the annealing of "hardness" in molybdenum polycrystalline aggregates ended at 1050°C.

Samples of oriented molybdenum single crystals were irradiated with neutrons in a nuclear reactor; the maximum intensity of the thermal neutron flux was 1 to $2 \cdot 10^{-13}$ neutrons/cm$^2 \cdot$sec, fast neutrons (over 1 MeV) contributing no less than 10% to the thermal neutron flux according to an approximate estimate. The integral flux (dose) for the period of irradiation of the samples (30 h) equalled $2 \cdot 10^{18}$ neutrons/cm^2; the sample temperature never exceeded 80°C. The microhardness before and after irradiation appears in Fig. 1. The number of defects arising in the annealed samples as a result of irradiation exceeded the number present before annealing, as indicated by the change in microhardness. Only for the (100) face was the microhardness slightly lower, and this may be explained by the high mobility of the defects on this face. In the case of the unannealed samples, the microhardness of the (110) face before and after irradiation differed little from the values corresponding to the irradiated and annealed sample. Hence above a certain limit it is difficult to create new defects on this most densely packed face.

The fact that the microhardness of the irradiated samples having their surfaces parallel to the (111) plane was very much the same for both annealed and unannealed samples once again supports the view that defects on the (111) planes were not annealed. For this face higher annealing temperatures are evidently required.

The most natural explanation for the radiation hardening of metals is that obstacles are created preventing the movement of dislocations [4]. The movement of dislocations may be prevented by vacancies and interstitial atoms, individual dislocations, groups of dislocations (networks, walls, loops), grain boundaries, twins, and the interfaces between blocks. In addition to this, more dislocations are created. The presence of blocks, i.e., disoriented parts of the same plane of the crystal, may be detected on the Laue photographs (Fig. 2). After irradiation, the spots on the Laue photographs are split, indicating the diminution and disorientation of the blocks.

References

1. Effect of Nuclear Radiations on Materials, Izd. AN SSSR, Moscow (1962).
2. Sh. Sh. Ibragimov and V. S. Lyashchenko, Fiz. Met. Metallov., 10(2):183.
3. G. N. Shuppe, Electron Emission of Metallic Crystals, Izd. SAGU, Tashkent (1957).
4. Mechanism of the Hardening of Solids, Izd. Metallurgiya, Moscow (1965).

KINETICS OF CATHODIC ADSORPTION PHENOMENA

Sh. A. Ablyaev and L. Ya. Alimova

The kinetics of gas adsorption by silicagels or zeolites subjected to the action of high-frequency discharges have still not been fully studied. It was shown earlier [1, 2] that there was a complete analogy between the effects of gamma rays and high-frequency discharges in creating surface radiation effects on solids. By using this analogy we may extend existing kinetic methods of studying radiation adsorption [3-5] to cathodic adsorption phenomena.

In this paper we shall attempt a kinetic analysis of the additional adsorption of hydrogen by the zeolite 13X (NaX) under the influence of a high-frequency discharge. As in the case of gamma-adsorption effects [6], the following division will be made in the kinetic processes:

a) Adsorption of hydrogen on the surface of a zeolite previously subjected to high-frequency discharges;

b) adsorption of hydrogen on a zeolite while a high-frequency discharge is actually acting.

These cases of induced adsorption are in many respects identical with the corresponding aspects of the gamma-adsorption effect; however, they have their own specific characteristics as well. Thus, in the case of a previously irradiated surface, in order to activate the surface by means of the high-frequency discharges, a certain amount of inert gas (He) must be first introduced into the adsorption space; this activates the surface during the discharge. The inert gas itself is only slightly adsorbed, occupying a small proportion of the active centers created. A special feature in the adsorption of hydrogen on the surface of a zeolite while a high-frequency discharge is actually acting upon the latter lies in the fact that the large quantity of gas required to produce a saturation effect must not be introduced into the adsorption space all at once, thus creating a relatively high gas pressure, since at such pressures the low-power discharge providing the positive adsorption effect cannot be ignited. Hence the required amount of gas must be introduced into the discharge space in successive steps as adsorption proceeds.

Adsorption of Hydrogen by Zeolite 13X (NaX) during the Operation of a High-Frequency Discharge. An adsorption ampoule with several side tubes containing hydrogen, itself containing a certain amount of finely divided zeolite 13X (NaX) and having a volume of 350 cm^3, was subjected to careful preliminary heat treatment (300°C, P = 760 mm Hg, t = 1 h) and vacuum heat treatment (350°C, P = 10^{-2} to 10^{-5} mm Hg, t = 8 to 10 h). Then the ampoule containing the zeolite was flushed with hydrogen and again evacuated to the limiting pressure of 10^{-5} to 10^{-6} mm Hg and filled with hydrogen (P = 10^{-1} mm Hg). The ampoule so prepared was subjected to the action of a high-frequency discharge. After a certain time, an additional amount of hydrogen was introduced from the side tubes, and irradiation continued until the next saturation was achieved, and so on. Operations continued in the same way until the hydrogen ceased being adsorbed on the zeolite under the influence of the discharge.

Adsorption of Hydrogen on the Surface of a Zeolite Previously Subjected to High-Frequency Discharges. It is well known [2] that a high-frequency discharge will not activate the surface of a zeolite with respect to inert gases (He, Ar). The induced adsorption of these two gases is almost zero. We used this property of zeolites and inert gases in order to make an experimental investigation into the adsorption of gases by zeolites after the high-frequency discharges acting on these had ceased.

An adsorption ampoule containing 4 g of finely divided zeolite 13X (NaX) and furnished with a manometric tube was subjected to normal vacuum heat treatment. The ampoule was flushed several times with helium and sealed off from the vacuum line with an inert-gas pressure of 10^{-2} mm Hg in the ampoule. The adsorption ampoule was also furnished with an additional volume of 300 cm^3 separated from the main one by a thin barrier. After an equilibrium gas pressure had been established, the ampoule was introduced into a high-frequency field and a discharge was ignited. Since the helium was hardly adsorbed at all in this process by the zeolite surface, the pressure in the ampoule remained constant. The new adsorption centers created by the high-frequency discharge on the surface of the zeolite partly disappeared as a result of recombination and partly remained free. After the action of the high-frequency discharge had ceased, the thin glass barrier separating the main volume containing the zeolite from the additional volume containing hydrogen (P = 300 mm Hg) was broken. The hydrogen came into contact with the surface of the zeolite activated by the discharge and the adsorption process commenced (Fig. 1).

When high-frequency discharges act upon zeolite surfaces, defects are created and these enter into the structure of the active adsorption centers. Experimental results confirm that the additional adsorption of hydrogen at the instant of the occurrence of a discharge is much greater than the adsorption of hydrogen on a previously irradiated zeolite surface.

Evidently the action of a high-frequency discharge on the surface of a solid creates both short- and long-lived adsorption centers. The short-lived centers, if not brought into contact with gas particles, vanish after a short time by way of recombination. Hence a previously irradiated surface only accommodates the long-lived centers.

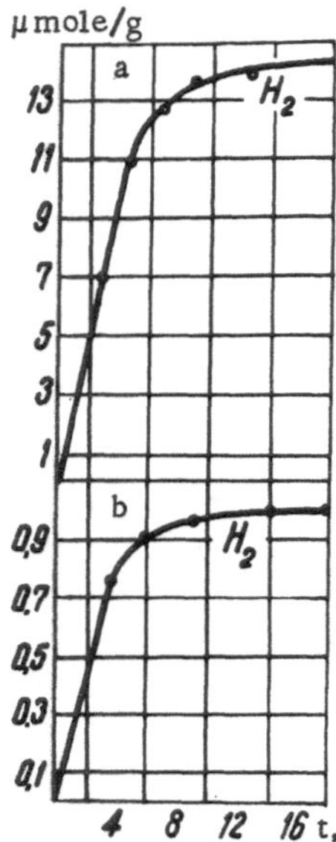

Fig. 1. Adsorption of hydrogen by zeolite 13X (NaX): a) during the action of a high-frequency discharge; b) on a previously irradiated surface.

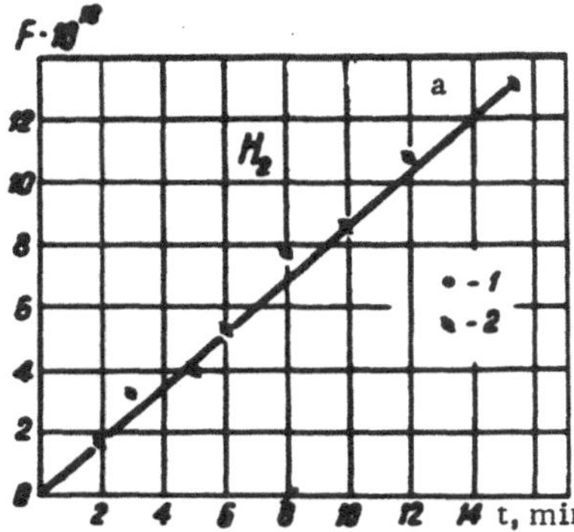

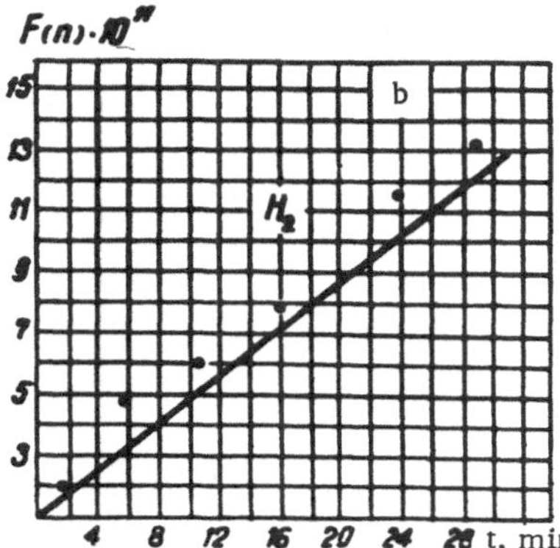

Fig. 2. Analysis of experimental results: a) by formula (2); b) by formula (5).

When gas particles come into contact with an irradiated surface, they are primarily adsorbed by such centers, and only after the capture of all these most active centers does the occupation of the remainder commence. Hence the velocity of the adsorption process on irradiated zeolites should be greater. This is in fact found in practice. The adsorption of hydrogen on a zeolite surface previously irradiated with a high-frequency discharge is identical with the adsorption of hydrogen on the surface of a gel subjected to gamma irradiation. Hence the kinetic equation for the adsorption of hydrogen on the surface of a zeolite previously subjected to a high-frequency discharge should be analogous to the corresponding kinetic equation representing the adsorption of hydrogen on the gamma-irradiated surface of silicagel [4, 5].

We may therefore consider that the rate of hydrogen adsorption on the previously irradiated surface of zeolite 13X (NaX) is proportional to the number of active centers created by the high-frequency discharge on the whole surface of the zeolite. In addition to this, it is proportional to the gas pressure, raised to a certain power equal to the order of the reaction with respect to hydrogen pressure.

If α-active centers are created on 1 cm^2 of zeolite surface by the high-frequency discharge, there will be $S\alpha$-active centers over the whole surface. Let us use Θ to denote the proportion of occupied centers; then the number of active centers on the surface at a certain instant of time t will be

$$S\alpha(1-\Theta)$$

The differential equation for the kinetics of hydrogen adsorption on the previously irradiated surface of zeolite 13X (NaX) is

$$-\frac{dn}{dt} = K_p n^{\nu} S\alpha(1-\Theta)$$

where K_p is a certain constant (a function of temperature),
ν is the order of the reaction with respect to the hydrogen pressure; in the present case $\nu = 1$.

Let n be the number of molecules in the gas phase at the instant of time t;
n_0 be the number of molecules in the gas phase at the instant of time $t = 0$.

Let us denote $S\alpha - n_0 = B$; then

$$-\frac{dn}{dt} = K_p n(B+n) \tag{1}$$

Integrating the resultant equation between the limits

$$t=0 \qquad n=n_0$$
$$t=t \qquad n=n$$

we have

$$\frac{1}{B}\left(\ln\frac{n_0}{n_0+B} - \ln\frac{n}{n+B}\right) = K_p t \tag{2}$$

A graphical comparison between solution (2), kinetic equation (1), and experimental results relating to the adsorption of hydrogen on the irradiated surface of the zeolite reveals excellent agreement, indicating the validity of the assumptions made in deriving Eq. (1). Figure 2a presents a graph of the experimental results analyzed on the basis of Eq. (2), the vertical axis representing the left-hand side of Eq. (2) and the horizontal axis the time.

In view of the fact that a reasonably perfect analogy exists between the effects of gamma rays and high-frequency discharges in creating surface radiation effects, the adsorption of hydrogen during the action of a high-frequency discharge should take place in the same way as in the presence of gamma irradiation.

The kinetics of hydrogen adsorption are thus described by the same equation as in the case of gamma irradiation [4, 5]

$$-\frac{dn}{dt} = K_p I \sqrt{n}\,(B_0 + 2n) \tag{3}$$

where I is the dose rate of high-frequency irradiation;

$$B_0 = S\alpha - 2n_0$$

Integrating this equation between the limits

$$t = 0 \qquad n = n_0$$
$$t = t \qquad n = n$$

we obtain

$$\frac{1}{\sqrt{2B_0}}\left(\ln\left|\frac{\sqrt{2n_0}+\sqrt{|B_0|}}{\sqrt{2n_0}-\sqrt{|B_0|}}\right| - \ln\left|\frac{\sqrt{2n}+\sqrt{|B_0|}}{\sqrt{2n}-\sqrt{|B_0|}}\right|\right) = K_p I t \tag{4}$$

$$(B_0 < 0)$$

$$\frac{1}{\sqrt{2B_0}}\left(\arctan\sqrt{\frac{2n_0}{B_0}} - \arctan\sqrt{\frac{2n}{B_0}}\right) = K_p I t \tag{5}$$

$$(B_0 > 0)$$

Comparison between the theoretical and experimental results indicates considerable deviations. We see from Fig. 2b, in which the vertical axis represents the left-hand side of Eq. (5), that the experimental points lie on the same side of the theoretical curve (in the direction of higher values) throughout the range studied.

In our own opinion the reason for this discrepancy is as follows. When hydrogen is adsorbed on the surface of the zeolite at the actual instant of the discharge, a fairly large quantity of gas is absorbed. The momentary introduction of such a large quantity of hydrogen into the discharge space creates a high pressure, preventing the ignition of a discharge of the intensity required. It was for this reason that the hydrogen was introduced into the discharge space in small portions. As a result of this the hydrogen pressure at which adsorption took place was always lower than the pressure which would have existed if all the hydrogen had been introduced at the same instant. Hence the adsorption rate and degree of adsorption always remained smaller, i.e., the experimental values appeared too low, and the corresponding curves appeared above the theoretical ones in the graph throughout the region under consideration.

Thus the kinetics of adsorption on a previously irradiated zeolite surface may be qualitatively described by the same differential equation as in the case of gamma rays.

However, the use of the differential equation governing the kinetics of hydrogen adsorption on silicagel at the instant of gamma-ray irradiation for the present case of adsorption on a zeolite during the operation of a high-frequency discharge yields results quantitatively not in agreement with experimental data. This is apparently because of the particular manner in which the experiments on the kinetics of adsorption processes were conducted.

References

1. S. V. Starodubtsev, Analogy between the radiative effects of radiation and slow charged-particle flows, in: Radiation Effects in Solids, Izd. AN Uzbek SSR, Tashkent (1963).
2. Sh. A. Ablyaev and S. V. Starodubtsev, Radiation Effects on the Surface of Gels, Izd. Uzbek SSR, Tashkent (1964).
3. Yu. A. Kolbanovskii and Yu. V. Pepelyaev, Kinetika i Kataliz, 6:237 (1965).
4. Yu. A. Kolbanovskii and M. N. Masterova, Kinetika i Kataliz, 7:727 (1966).
5. G. M. Dolidze, Yu. A. Kolbanovskii, and L. S. Polak, Kinetika i Kataliz, 6:897 (1965).

MASS-SPECTROMETER DETERMINATION OF TRACES OF ALKALI ELEMENTS IN SILICON

Sh. A. Ablyaev and L. Ya. Alimova

The study and elucidation of the part played by impurities in semiconducting materials is of major importance in semiconductor technology. There are various analytical methods of determining impurities in solids; one convenient means is mass-spectrometer analysis, using the phenomenon of surface ionization. The surface ionization of alkali elements on a heated silicon surface has been studied on a small number of occasions [1-4] and the corresponding ion emission determined.

In this paper we shall use the phenomenon of the intrinsic emission of silicon in order to determine alkali-metal impurities and study the kinetics of their evolution at high temperatures. The experiments were carried out on single-crystal silicon of the n type. In studying the impurities in the silicon we used an MI-1305 mass spectrometer, the tungsten strip in the strip source of the instrument being replaced with a silicon plate 17 × 2.4 × 0.5 mm in size. The working vacuum in the analyzer chamber was 5 to $7 \cdot 10^{-7}$ mm Hg. The sample temperature was determined by a method described in another paper [5].

Figure 1 shows the temperature dependence of the evolution of various alkali elements from silicon. We see from the form of the curves that substantial ionization of potassium,

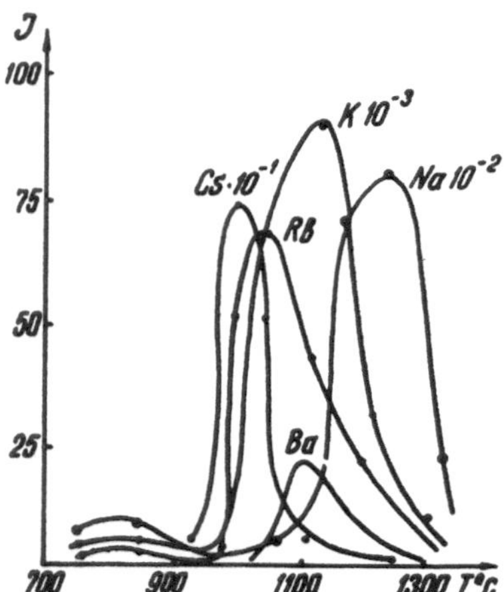

Fig. 1. Evolution of various alkali elements from silicon at various temperatures.

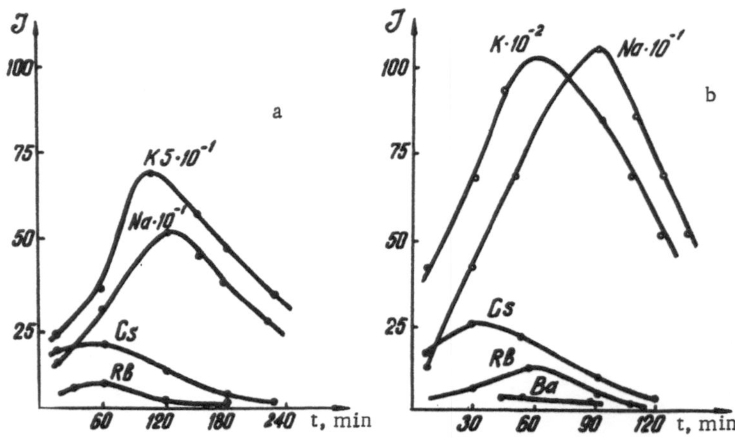

Fig. 2. Kinetics of the evaporation of alkali impurities from silicon at 1050 (a) and 1200°C (b).

rubidium, and sodium begins at 700°C; after passing through a flat maximum, it starts falling and reaches a minimum at sample temperatures of about 900°C.

On further raising the sample temperature, there is a sharp increase in ion emission, since the diffusion coefficients and ionization probability rise exponentially with increasing sample temperature.

The ion currents corresponding to individual elements reach their maximum values at 1050, 1080, 1100, and 1200°C for cesium, rubidium, barium (and potassium), and sodium, respectively, and then start falling rapidly with any further increase in the temperature. This sharp fall in the ion current is evidently due to a reduction in the amount of impurity in the sample and a certain fall in the work function of the emitter itself.

Figure 2 presents the kinetic curves corresponding to the revelation of various impurities at 1050 and 1200°C. The curve maxima corresponding to sodium and potassium ions at a temperature of 1050°C appear two hours after this temperature is first attained (Fig. 2a). The barium ions are not recorded at all, despite attempts at detecting them over a period of several hours. The positive emission current at 1200°C is much greater than at 1050°C (Fig. 2b). The emission maxima of the Na^+ and K^+ ions appear one hour after initial heating of the silicon sample, i.e., an hour earlier than at 1050°C.

References

1. N. G. Ban'kovskii and B. N. Formozov, Izv. Akad. Nauk SSSR, Ser. Fiz., 28:2048 (1964).
2. É. Ya. Zandberg and V. I. Paleev, Zh. Tekhn. Fiz., 35:1308 (1965).
3. É. Ya. Zandberg and V. I. Paleev, Zh. Tekhn. Fiz., 35:2092 (1965).
4. É. Ya. Zandberg and V. I. Paleev, Zh. Tekhn. Fiz., 34:2048 (1964).
5. L. M. Gol'shtein, Dokl. Akad. Nauk SSSR, Vol. 158, No. 2 (1964).

STUDY OF THE ADSORPTION OF GAS MIXTURES ON THE IRRADIATED SURFACES OF SILICAGELS BY THE METHOD OF INFRARED SPECTROSCOPY

Sh. A. Ablyaev, L. Ya. Alimova, and Z. S. Settarova

The effect of radiation on the adsorption of gas mixtures by silicagels has been little studied up to the present time. Mixtures in which one or several components are chemisorbed have not been studied at all. During the gamma irradiation of silicagels and zeolites, the components of a binary gas mixture (O_2, N_2) are adsorbed in accordance with their partial pressures [1] on the radiation centers, quite independently of one another. On studying two-component (H_2, O_2) (H_2, N_2) or three-component (H_2, O_2, N_2) mixtures, the degree of adsorption is found to depend very considerably on the nature of the components. The presence of hydrogen in the adsorption space increases the adsorption of oxygen by up to 20% and that of nitrogen by up to 6%. Our own experiments showed that analogous results followed the subjection of silicagels to high-frequency discharges.

The additional increase in the adsorption of oxygen and nitrogen on the surface of silicagel subjected to either gamma irradiation or high-frequency discharges in the presence of hydrogen may be explained by the fact that irradiation promotes the formation of chemical compounds between the hydrogen atoms and oxygen, nitrogen, or the constituent elements of the silicagel on the surface of the latter.

The presence of chemical compounds of the gas-mixture components with each other and with the constitutent elements of the adsorbent itself on the surface of irradiated silicagels was established by a method based on infrared spectroscopy. The infrared absorption spectra of the silicagels were studied before and after the adsorption of gas mixtures and also after the action of gamma rays or high-frequency discharges. For this purpose a suitable vacuum adsorption cuvette was developed, samples being heat-treated in this at a temperature of up to 320°C at a vacuum of 10^{-4} to 10^{-5} mm Hg.

The vacuum adsorption cuvette (Fig. 1) consisted of a T-shaped glass ampoule-cuvette furnished with two plane-parallel windows C made of calcium fluoride (CaF_2) with a spectral transmission range of 0.5 to 9 μ (96%) and an absorption band at 9 to 12 μ. The CaF_2 windows were attached to the cuvette with epoxy resin. A mobile metal construction ADB was placed in the adsorption cuvette; this comprised an iron cylinder A, two flat rings B with pressure screws, and a holder D. Between the two flat rings a round plate of barium fluoride with a spectral transmission range of 0.5 to 12 μ (96%) and an absorption band at 12 to 15 μ was pressed. On one side of the BaF_2 plate a fine powder of KSK silicagel some $5 \cdot 10^{-4}$ g/cm^2 thick was deposited.

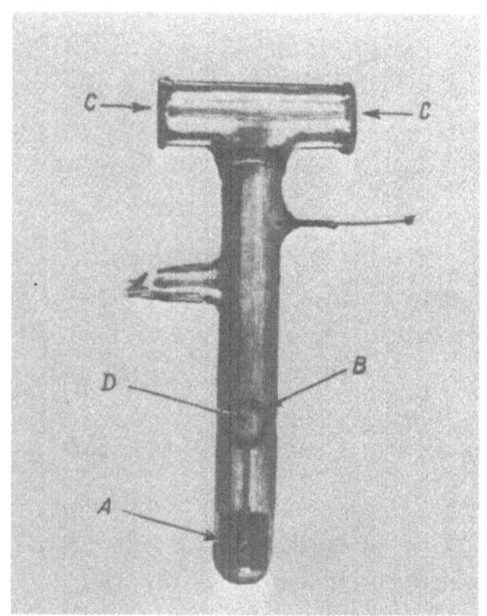

Fig. 1. Vacuum adsorption cuvette.

The devices prepared in this manner were subjected to careful vacuum heat treatment at 300°C for 10 to 12 h. The metal cylinder with the gel sample lay in the lower part of the vertical tube. After vacuum heat treatment, one of the devices was sealed off from the vacuum line, a gas pressure of the order of 10^{-5} mm Hg remaining within it. Before sealing the other device, specific quantities (1 cm^3) of hydrogen, oxygen, and nitrogen were passed into it, the total gas pressure being 9 to 10 cm Hg. Then the infrared absorption spectra of the silicagels in the two cuvettes were recorded by means of an ISK-14 spectrophotometer.

In recording the infrared spectra, the sample was moved from the lower part of the tube into the space between the parallel CaF_2 windows by means of an electromagnet. In one case a device of this kind with a gas mixture inside it was subjected to the action of gamma rays (integral dose 2.5 to 3 million R) and after the irradiation the infrared absorption spectrum of the silicagel was recorded again. In the other case the gas was pumped out after recording the infrared spectrum, and the surface of the silicagel was activated with a high-frequency discharge; then the same amounts of gas as before were introduced and the spectrum was recorded again from the silicagel. These infrared absorption spectra of KSK silicagel plotted before (a) and after (b) gamma irradiation or after the action of a high-frequency discharge are shown in Fig. 2. Spectra I, II, and III relate to nonirradiated gel: I and II were obtained before and after vacuum heat treatment, III after the successive introduction of H_2, O_2, N_2 into the cuvette. The spectrum of the untreated gel shows the following absorption bands: 3750 cm^{-1}, 3650 cm^{-1}, 3600 cm^{-1}, 3280 cm^{-1}, 3200 cm^{-1}, 2500 cm^{-1}, 1630 cm^{-1}, 1530 cm^{-1}, 1425 cm^{-1}. As a result of heat treatment, certain changes occur in the infrared spectrum, and the absorption band at 1630 cm^{-1} disappears almost completely.

The introduction of gases after vacuum heat treatment causes a slight intensification of the bands lying between 3650 and 3520 cm^{-1}, and the band at 1630 cm^{-1} again becomes appre-

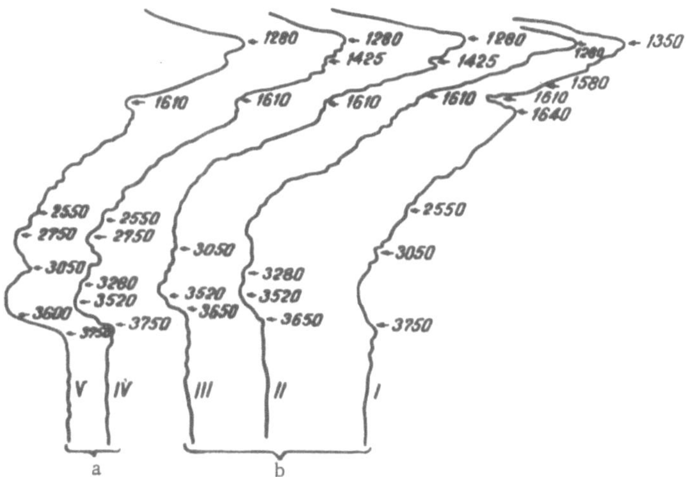

Fig. 2. Infrared absorption spectra, cm^{-1}: a) before irradiation; b) after gamma irradiation or the action of a high-frequency discharge.

ciable. The bands lying between 3704 and 3570 cm^{-1} belong to the independent OH structural groups weakly coupled to the other atoms in the crystal lattice of the adsorbent. Interaction between the hydroxyl groups, arising as the latter come together under the influence of a hydrogen bond, leads to the weakening of the coupling between the hydroxyl hydrogen and oxygen atoms and to the displacement of the corresponding absorption band in the direction of longer wavelengths in the range 3700 to 3225 cm^{-1}.

The range 2750 to 2550 cm^{-1} accommodates the valence and deformation vibrations of the C=H groups, and the range around 1610 cm^{-1} accommodates the N–H deformation vibrations [2, 3]. Irradiation of the silicagel does not result in the appearance of any new absorption bands (spectra IV and V), although the bands existing in the spectrum of the nonirradiated gel are considerably strengthened. The absorption bands in the range 3650 to 3050 cm^{-1} are strengthened by a factor of very nearly two. The absorption bands in the range 2750 to 2550 cm^{-1} become more noticeable.

Under γ irradiation or the effects of a high-frequency discharge, complicated radiochemical processes occur on the surface of the silicagel; as a result of these processes, not only free and coupled OH groups but also surface chemical compounds of nitrogen with hydrogen containing N–H groups (valence vibrations between 3500 and 3300 and deformation vibrations between 1650 and 1580 cm^{-1}), charged NH_3^+ groups (valence and deformation vibrations in the ranges 3380 to 3280 and 3350 to 3150 cm^{-1}), and NH_2^+ groups (valence vibrations at 2700 cm^{-1}, deformation vibrations between 1620 and 1560 cm^{-1}) may appear. On introducing air up to atmospheric pressure into the cuvette (spectrum V), the absorption bands at 3650 to 3280 and 1630 cm^{-1} are still further intensified.

Thus the additional adsorption of oxygen and nitrogen arising from the presence of hydrogen is due to the development of free and combined OH groups and compounds of hydrogen with nitrogen forming N–H, NH_3^+, NH_2^+ groups on the surface of the silicagel.

References

1. Sh. A. Ablyaev and S. V. Starodubtsev, Radiation Effects on Gel Surfaces, Izd. AN Uzbek SSR, Tashkent (1964).
2. K. Lowson, Infrared Absorption Spectra of Inorganic Substances [Russian translation], Izd. Mir, Moscow (1964).
3. L. Bellami, Infrared Spectra of Molecules [Russian translation], IL, Moscow (1957).

STUDY OF CERTAIN MACROSCOPIC DEFECTS IN REFRACTORY METALS

G. A. Klein and V. A. Krakhmalev

All crystalline materials, including metals, contain a wide variety of crystal–structural defects such as vacancies, interstitial atoms, lines of dislocations, grain boundaries, subsidiary boundaries of mosaic blocks, pores, cracks, precipitated impurities, and so on. Many of the properties of metal crystals are directly associated with these defects [1-4]. A study of the various structural defects in metals is accordingly of major scientific and practical interest. Defects are most frequently studied in low-melting-point metals, steels, and alloys. The crystal–structural defects of the refractory metals, molybdenum, tungsten, rhenium, etc., have been little studied, although these metals have the most promising characteristics in current technology. We therefore decided to study pure molybdenum and molybdenum–zirconium alloys obtained by electron-beam vacuum zone melting and also thorium-containing tungsten (wire of the VT-10 type 0.2 mm in diameter).

§1. Method of Single-Stage Replicas for the Electron-Microscope Study of Macroscopic Defects in Refractory Metals

In order to study massive metal samples in the transmission electron microscope, the method of replicas or impressions [5] is employed; in this method it is not the actual sample which is examined in the electron microscope, the sample being completely opaque to electrons as a result of its thickness, but a thin film of some suitable substance deposited on the sample and then removed. This replica (composed of carbon or plastic, for example) reproduces the microstructure of the massive metal sample to a high degree of accuracy and enables it to be studied at high magnifications.

We developed a method of single-stage replicas, simple and convenient for the rapid examination of polished surfaces, pores, and impurity precipitates at intercrystallite boundaries [6].

The replicas may be studied in the electron microscope immediately after being deposited. In order to separate the replicas from the sample there is no need to use strengthening layers, nor chemical or electrolytic etching, such as would damage the sample itself.

Replicas of intercrystallite fractures and the polished faces of refractory metals were obtained by the thermal vacuum deposition of carbon in the manner generally accepted [15]. The optimum thickness of the replicas was determined experimentally. By way of a criterion indicating the right moment to stop depositing carbon, we took pieces of transparent plexiglass and placed these alongside the object. The deposition was halted when a film of golden or chestnut color appeared on the plexiglass. The sample was taken from the vacuum apparatus, and, with the replica on top, slowly immersed in distilled water at an angle of 30 to 45° to the surface of the latter.

A purely carbon replica suffers no wetting by water; thus as the sample is immersed in the water the deposited carbon layer floats off and remains on the surface. By using this method of separating the replicas, the preparation of a particular metal sample takes just 10 min. Replicas obtained by the thermal vacuum evaporation of substances other than carbon are also easily removed from the polished faces and intercrystallite fractures of refractory metals.

In order to study cleavages and etched microsections of metals having a well-developed relief, single-stage replicas removed by means of a gelatine layer may be employed as well. The time required to secure such sample replicas is rather longer (order of a day).

§2. Negative Crystals (Pores) in Molybdenum and Their Surface Structure

Experimental observations of the behavior, shape, and contour characteristics of pores in crystalline materials are usually based on the use of an optical microscope [7]. The transmission study of thin metal foils in the electron microscope enables us to determine the general features of very small pores and also the influence of various factors on their density and size [8, 9]. The use of replicas for observing and studying macroscopic defects in metals enjoys great advantages over both optical and transmission electron microscopy. By using replicas we may study both very small pores (within the limits of the electron-microscope resolving power) and also large pores such as may even be seen in the optical microscope. In addition to this, it is only by the replica method that we can obtain a general idea of the structure of pore (negative-crystal) surfaces in metals.

In single crystals of electron-beam zone-melted molybdenum, pores occur both inside and on the surface of the crystals. An electron-microscope study of replicas taken from pores of pure and zirconium-alloyed molybdenum showed that the surfaces of the negative crystals had a peculiar fine structure not seen under the optical microscope.

The objects for electron-microscope examination were fresh, unetched trans- and intercrystallite surfaces of brittle cleavages and fractures of pure and zirconium-alloyed (0.15% Zr) molybdenum single crystals obtained by electron-beam zone-melting. Single-stage carbon and platinum–carbon replicas were separated from the transcrystallite brittle cleavages of the single crystal by means of a layer of gelatin, and from the surfaces of the intercrystallite fractures by immersing the samples in cold distilled water [6]. Judging by the epigrams thus obtained, the transcrystallite cleavages of the molybdenum single crystals coincided with the {001} and sometimes with the {110} planes. Pores in these samples included some which were sharply bounded by crystal faces (Fig. 1a) and others of an intermediate shape between a polygon and a sphere (Fig. 1b). In the single-stage replica method, with the carbon deposited at an acute angle to the surface of the cleavage, the pores were classified by reference to the shadows lying within their geometrical contour. A special feature of the pores shown in Fig. 1a is the existence of a fine face structure, in the form of a system of concentrically-arranged square steps around the exit points of the [001] direction. The edges of the steps were directed at an angle of 45° to the [010] and [100]. Clearly these steps are bounded by planes close to the (110) and (001).

The molybdenum pores (negative crystals) of intermediate form (Fig. 1b) are characterized by a mirror-symmetrical surface structure, consisting of a set of strips or bands, each

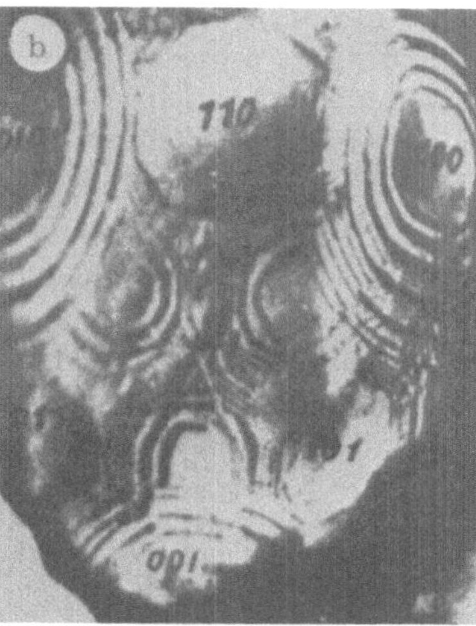

Fig. 1. Electron micrographs of replicas from brittle transcrystallite cleavages of molybdenum: a) platinum–carbon replica from a cleavage plane (100), magnification 7500; b) carbon replica from a cleavage plane (110), magnification 25,000.

of these being bordered by a system of concentric polygonal steps. The structure of the pore surface as a whole is symmetrical relative to the [110] direction (in Fig. 1b half the pore surface is shown). In the transcrystallite brittle cleavages of molybdenum, pores are encountered very rarely and are therefore difficult to observe. In the intercrystallite fracture surfaces of pure molybdenum and molybdenum containing 0.15% zirconium, these pores are far more numerous; they form disordered aggregates, lattices, and chains. The small pores (under 1 μ) have no fine face structure (within the limits of the resolution of the EM-5 electron microscope), but have the shape of polygons. The surface of the large pores (usually spherical or ellipsoidal in shape) consists of a set of poles with a system of concentric polygonal terraces (Fig. 2). Sometimes the steps bordering the bands have an internal complex structure, consisting of finer steps.

Fig. 2. Electron micrograph of a platinum–carbon replica from an intercrystallite fracture of molybdenum. Magnification 30,000.

The formation of the peculiar stepped structure of systems of concentric polygonal steps around a plane face on the surface of molybdenum pores (negative crystals) is evidently due to the equilibrium form of the surface developing during crystallization.

The surface structure of the large pores (Fig. 2) studied by the replica method has much in common with the images obtained from single-crystal sharp-pointed filaments of refractory metals (molybdenum, tungsten, rhenium, etc.) in electron [10] and ion [11] micro-projectors (projection microscopes). The existence of concentric steps around the exit points of certain crystallographic directions on the equilibrium surfaces of the negative crystal constitutes a special kind of visualization of the emission pictures revealing the fine structure of the emitters, which clearly arises from their heat treatment. Hence the fine structure of the bright places in the emission pictures of certain sharp filaments [12] may be explained by the high resolving power of the instrument.

The fine surface structure of the molybdenum pores (negative crystals) confirms the fact that in equilibrium the crystal surface should have a stepped structure [13, 14]. The flat parts of the crystals constituting the stepped structure of the equilibrium surface should (as indicated by L. D. Landau [13]) be the smaller, the greater the indices of these faces. We see from Fig. 1 that the $\{100\}$ faces have the greatest dimensions.

§3. Morphological Forms of Impurity Precipitates on the Intercrystallite Surfaces of Molybdenum

The wide use of molybdenum as a construction material is impeded by its low-temperature brittleness. The purity of the molybdenum has a considerable effect on the stress at which brittle failure sets in. It was shown earlier [16] that the introduction of 0.0001 to 0.0002% of oxygen into molybdenum causes brittleness even at room temperature. The ductility of molybdenum in the cold state may also be reduced by nitrogen and carbon. At low and room temperatures, polycrystalline molybdenum ruptures along the grain boundaries on which the impurities have precipitated. In metallographic examinations of ordinary microsections, it is almost impossible to observe impurity precipitates at the crystallite boundaries of molybdenum and to determine their chemical composition and morphological form. This is explained by the fact that the impurity precipitates on the crystallite boundaries of molybdenum are so small in thickness as to be beyond the limits of the resolving power of the optical microscope. At the same time, the development of new phase precipitates as a result of the selective segregation of impurity atoms on the intercrystallite surfaces of separation of molybdenum constitutes one reason for the sharp changes in physicomechanical properties.

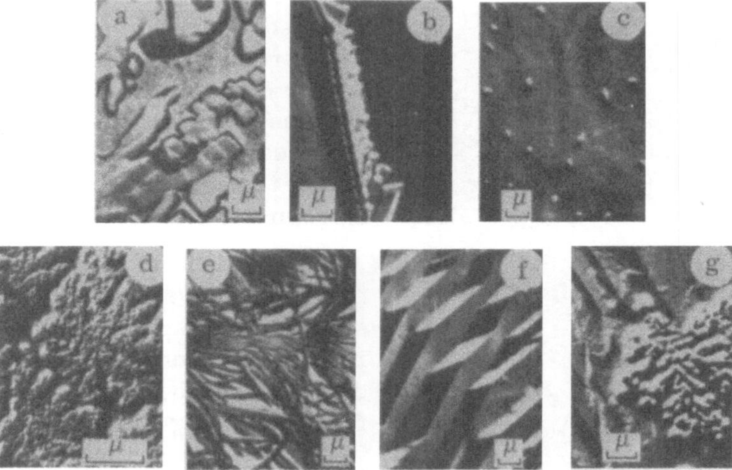

Fig. 3. Electron micrographs of replicas from different morphological forms of impurity precipitates on the intercrystallite interfaces of molybdenum: a) 11,400; b) 4000; c) 28,500; d) 12,600; e) 7000; f) 6000; g) 6000 magnification.

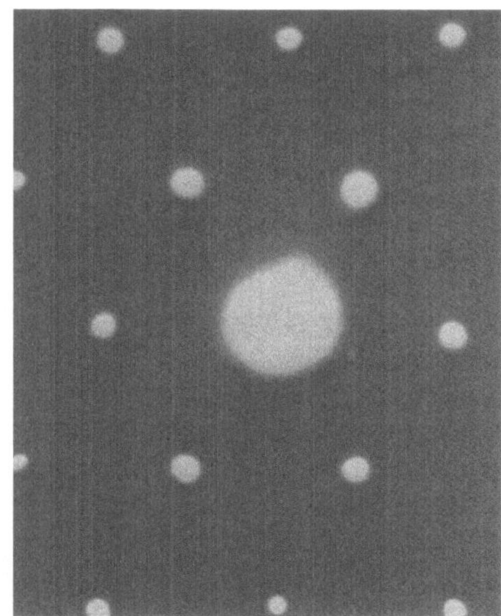

Fig. 4. Electron microdiffraction pattern from impurity particles of the first type.

Little work has been carried out on the morphological forms of the impurity precipitates at the intercrystallite boundaries of molybdenum [16, 17].

Using the replica technique, we studied the upper ends of single-crystal molybdenum bars grown by electron-beam zone melting. After splitting off the polycrystalline end to reveal the fresh surface of the molybdenum single crystal formerly attached to it, we vacuum-deposited a single-stage carbon replica in the same place. In order to give contrast to the images of the precipitating phase, the intercrystallite surface of the fracture was shadowed with gold. The replicas were detached from the test surfaces by immersion in water [6]. The impurities which had precipitated at the crystallite boundaries in no way impeded the removal of the replica.

We inspected a large number of replicas obtained from the intercrystallite surfaces of a molybdenum fracture in the electron microscope. In morphological form, the impurity precipitates may be divided into three main types. The first, most frequently-encountered type of impurity precipitates is represented by plane, two-dimensional (to a first approximation) precipitates in the form of squares, rectangles, and triangles (Fig. 3a). Precipitates of this type present in a high density are capable of forming conglomerates in the form of large, discontinuous films and also precipitates of very fantastic shapes (Fig. 3b) extending for a long way in one of the crystallographic directions.

Precipitates of the second type are globular (Fig. 3c), their equilibrium form being reminiscent of a grain of wheat 1000 to 2500 Å in size; they are found more rarely than the first type. In certain parts of the intercrystallite space they form continuous films composed of spherulitic grains (Fig. 3d) and also relief figures reminiscent of leaf structures. The surface of such formations is granular.

Most rarely of all we encounter impurity precipitates of the third type, in the form of long needles or sticks, sometimes forming continuous films (Fig. 3e).

The selective segregation of impurity atoms on the boundaries of crystallites leads to the embrittlement of not only the intercrystallite surfaces but also the regions of molybdenum lying close to these. We found cases in which the fracture passed, not exactly through the boundaries of the crystallites, but close to them. With this kind of crack propagation, a peculiar rupture pattern, reminiscent of the transcrystallite cleavages of molybdenum, appears (Fig. 3f); partly visible on the steps of the cleavage are precipitates of rounded or dendritic appearance (Fig. 3g).

By etching the intercrystallite fracture surface, some of the impurity precipitates were transferred to the replica (this mainly applied to those of the first type). Interpretation of the microelectron-diffraction patterns (Fig. 4) obtained from the extracted precipitates showed these to be MoO_2 single crystallites. The chemical composition of the impurity precipitates of the first type (the most numerous and most frequently encountered) confirms the fact that oxygen is almost insoluble in molybdenum. On exceeding the solubility limit it segregates at grain boundaries, forming oxides.

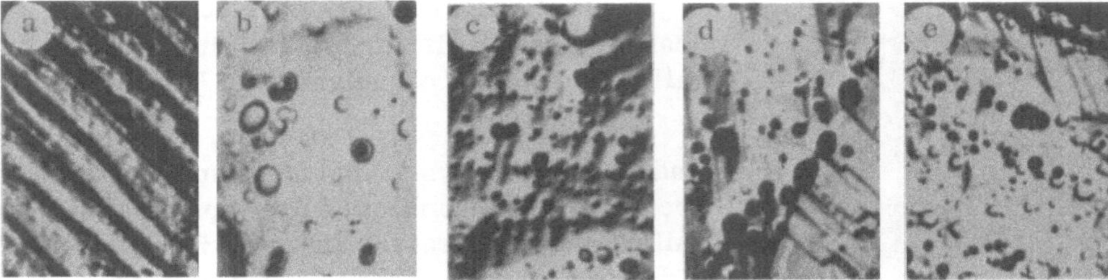

Fig. 5. Electron micrographs of replicas from etched longitudinal sections of a thorium-containing tungsten wire.

Thus an electron-microscope examination of the intercrystallite fracture surfaces may serve as one of the most sensitive methods of observing and chemically analyzing small amounts of precipitated impurities.

§4. Dislocation Origin of Precipitates in Aged Thorium-Containing Tungsten Wire

The segregation of impurity atoms on dislocations and their precipitation in the form of well-dispersed particles after certain forms of treatment has been successfully used in the direct observation of dislocation lines in optically-transparent crystals or those transparent to the infrared spectrum [18]. The revelation of the dislocation structure of massive samples of metals and alloys presents considerable experimental difficulties.

For certain alloys, mainly aluminum alloys with a few percent of copper, the method of oxide replicas has been used [19, 20] to study the role of dislocations in forming impurity precipitates and hence the dislocation structure itself. The new phase particles are in fact preferentially generated at dislocations, slip planes, and subsidiary boundaries. It is of particular interest in this connection to secure data relating to crystal-structural defects such as dislocations and impurity precipitates in refractory metals.

The object for our electron-microscope investigation was a thorium-containing tungsten wire 0.2 mm in diameter (type VT-10) aged by prolonged annealing (approximately 1000 h at 2400 to 2500°C in a vacuum of $1 \cdot 10^{-7}$ mm Hg). We prepared longitudinal microsections of the wire, etching this in an alkaline solution of potassium ferricyanide, carefully washing, and then drying. The etched surface of the longitudinal microsection of the wire was studied by the single-stage carbon-replica method [21] in the electron microscope.

An electron micrograph of the structure of an etched, unannealed longitudinal section of a thorium-containing tungsten wire is shown in Fig. 5a so as to provide a comparison standard when considering the changes taking place in the structure and defects of the wire while aging. As annealing continues, the sinuous structure of the wire becomes polycrystalline. In the main body of the grain, small spherical particles of impurity precipitates appear (Fig. 5b). The extracted impurity particles are given a strong contrast by virtue of the total absorption and scattering of the electrons in the corresponding material. Wherever no precipitated particles are captured, holes appear in the replicas, representing the impressions arising from the particles. Around some of these particles (Fig. 5b) concentric circles appear, looking gray on the background; the origin of these lies in the more intensive etching which takes place around the precipitates. We see from the same micrograph that some of the impurity particles consist of two and even three closely-adjacent minute particles.

As aging continues, the number of impurity precipitates in the wire increases sharply (Fig. 5c) and at the same time the diameter of some of the particles increases. Particularly large new phase precipitates are observed along the boundaries of adjacent grains (Fig. 5d). This micrograph clearly shows the white "shadows" cast behind the impurity particles captured by the replica, confirming that these constitute projections on the surface of the microsection.

Whereas, at the onset of aging, the arrangement of the dispersed particles was more or less random (Fig. 5b), as aging continues there is a preferential precipitation of new phase particles along certain lines parallel to the axis of the wire (Fig. 5c). Sometimes the precipitated particles are arranged along these preferential lines in double rows, as indicated in Fig. 5e, which shows the edge of the wire microsection.

One notices the fact that the precipitated particles are separated by specific intervals along these lines; the characteristic distance between these precipitates on the micrographs is several thousand angstroms. In an earlier treatment [22] this distribution of the small impurity particles was taken as constituting a good reason for assuming that each precipitate corresponded to one particular dislocation. If one also supposes that a minimum of one dislocation corresponds to each observed precipitate, then the dislocation density calculated from, for example, Fig. 5c approximately equals $5 \cdot 10^8$ cm^{-2}.

Thus, as a result of the preferential formation of highly-dispersed impurity particles on the dislocations, extremely valuable data may be secured as to the distribution and behavior of imperfections in the crystal lattice as a result of thermal aging.

References

1. R. M. Barrer, Diffusion in Solids [Russian translations], IL, Moscow (1948).
2. A. H. Cottrell, Dislocations and Plastic Flow in Crystals [Russian translation], Metallurgizdat, Moscow (1958).
3. V. S. Ivanova et al., Role of Dislocations in the Hardening and Softening of Metals, Izd. Nauka, Moscow (1965).
4. M. G. Lozinskii, Structure and Properties of Metals and Alloys at High Temperatures, Metallurgizdat, Moscow (1963).
5. A. N. Pilyankevich, Practice of Electron Microscopy, Mashgiz, Moscow (1961).
6. G. A. Klein and V. A. Krakhmalev, Method of single-stage replicas for studying intercrystallite surfaces of refractory metals, Zavod. Lab., 3:308 (1966).
7. Ya. E. Geguzin, Macroscopic Defects in Metals, Metallurgizdat, Moscow (1962).
8. R. S. Nelson, D. J. Mazey, and R. S. Barnes, Phil. Mag., 11:(109):91-111 (1965).
9. Michio Kiritani, J. Phys. Soc. Japan, 19(5):618-631 (1964).
10. G. N. Shuppe, Electron Emission of Metal Crystals, Izd. SAGU, Tashkent (1959).
11. E. W. Müller, Advances in Electronics and Electron Physics, Vol. 13 (1960), pp. 83-179.
12. H. Birkenschenkel, R. Haefer, and P. Mezger, Acta Phys. Auster., 7:402 (1953).
13. L. D. Landau, Collection Devoted to the 70th Anniversary of Academician A. F. Ioffe, Izd. SSSR, Moscow (1950), pp. 44-49.
14. I. M. Lifshits and A. A. Chernov, Kristallografiya, 4(5):788-791 (1959).
15. D. E. Bradley, Brit. J. Appl. Phys., 5:96 (1954).
16. L. E. Olds and G. W. P. Rengstorff, J. Metals, Vol. 8, No. 2, Sec. 2, pp. 150-155 (1956).
17. Yu. B. Malevskii, Electron Microscopy in Industry, Izd. Naukova Dumka, Kiev (1964).
18. Dislocations and Mechanical Properties of Crystals [Russian translation], IL, Moscow (1960).
19. V. G. Rakin and N. N. Buinov, Fiz. Met. Metallov., 18(6):877-887 (1964).
20. V. G. Rakin and N. N. Buinov, Dokl. Akad. Nauk SSSR, 121:(2):271-273 (1958).
21. G. A. Klein, V. A. Krakhmalev, V. P. Sokolova, and A. N. Filippov, Dokl. Akad. Nauk Uzbek SSR, 1:24-27 (1965).
22. R. Castaing, Comptes Rendus, 228:1341 (1949); J. Nutting and J. Thomas, Internat. Conf. on Electron Microscopy, London (1954).

PRODUCTION AND PROPERTIES OF ORIENTED SINGLE CRYSTALS OBTAINED FROM MOLYBDENUM POWDER

G. A. Klein, S. M. Mikhailov, L. Kh. Osipova, V. E. Grakov, G. N. Grishkov, and M. G. Sultanova

A method of growing oriented single crystals of refractory metals by electron-beam vacuum zone melting from compacted moldings was described earlier [1-3] and some of the physicomechanical and other properties of the single crystals in question were investigated.

In this paper we shall present the results of an investigation into the physicomechanical properties of oriented single crystals obtained from molybdenum powder after pressing, annealing in hydrogen, and electron-beam zone melting, involving the growth of a single-crystal bar based on a seed oriented in a specific crystallographic direction.

The original material was molybdenum powder with the following impurity content (wt. %): O_2 – 0.25, R_2O_3 – 0.03, Ni – 0.008, SiO_2 – 0.03, CaO_2 and MgO_2 – 0.008. The powder was mixed in alcohol (4 cm^3 to 1 kg of mixture) and without drying pressed into molds of circular cross section, using a press with a pressure of 2.5 to 3.2 tons/cm^2. The length of the pressed rod was 380 mm and the diameter 24 mm.

The pressed rods were placed in a nickel boat and sintered for 45 to 60 min in the furnace in a hydrogen medium at 1250°C. The sintered rods had a density of 65% of that of the cast metal. Electron-beam zone melting leading to the growth of oriented single crystals from the sintered rods was carried out in an electron vacuum furnace. For this purpose, the sintered rods were placed in cylindrical holders made from forged molybdenum rods 10 mm in diameter. In order to ensure the stability of the molybdenum rod, its ends were machined into the form of a cone; then a recess was made and the holder was fitted into this. The cone and recess at the ends enabled these to be sited accurately and reliably in the furnaces and also prevented the splashing of the metal when the first molten zone was formed at the conical end of the rod.

In order to place the rod in the zone-melting furnace, the supports were first attached to the upper and lower clamps, then the rod was inserted and pressed tightly against the ends. On reaching a vacuum of $5 \cdot 10^{-5}$ mm Hg the support-fixing points were sealed to the rod by means of the electron beam, after which the zone was created in the lower conical part of the rod and melting took place from bottom to top, the upper part of the rod over the zone being rotated at 100 rpm.

After the first zone pass at a rate of 4 mm/min, in which the rod was melted throughout its whole length, a single crystal with an orientation lying within the [100]—[110]—[111] stereographic triangle was formed; in the second pass this orientation of the single crystal was preserved even for zone velocities of 2 to 8 mm/min.

In order to grow single crystals with a specified crystallographic orientation, seeds in the form of cylinders 25 to 30 mm long were prepared from selected single crystals. The orientation of the axis was taken close to the [100], [110], [111], [112] directions and the seed plane was adjusted to the desired direction to an accuracy of 2°.

In every case the orientation was determined by the Laue back-reflection x-ray method. The single crystal grown after the first pass was cut into two parts near the lower support, the seed being placed in the gap and firmly pressed against the parts of the severed single crystal.

The second pass of the zone started from the seed at a rate of 2 mm/min along the single-crystal bar. The growing single crystal acquired the orientation of the seed along its whole length. It was found that the mean deviation of the single-crystal axis along the whole length of the bar (measured for a group of 100 oriented molybdenum single crystals grown in the manner indicated) never exceeded 2 to 4°. The deviation of the single-crystal axis from the specified orientation largely depended on the accuracy with which the seed was set in place.

Using the method just described, a series of molybdenum single crystals (bars) was grown in four crystallographic directions: [100], [110], [111], [112]. For testing the strength and ductility, cylindrical samples were turned from the bars, with a calculated (working) length of 100 mm and a diameter of 7 mm. In order to remove internal stresses, the samples were annealed at 1500°C for 2 h in a vacuum of $5 \cdot 10^{-5}$ mmHg. Samples were ruptured on a universal screw-type machine of the UM-5 type designed for tensile and compressive tests, incorporating an automatic recorder.

In the course of the tests the stress/strain diagrams were recorded in coordinates of load (kg) and elongation (mm) and the principal rupture characteristics were determined from these. In addition to this, the nature of the rupture taking place in molybdenum single crystals oriented in different directions was analyzed. The microhardness of the molybdenum single crystals was tested on the PMT-3 hardness tester, using a 100-g load and a loading period of 15 sec. For this purpose the microsections were polished mechanically and electrolytically in a mixture of sulfuric acid (25%), hydrochloric acid (12.5%), and methyl alcohol (62.5%) at a voltage of 30 to 40 V and a current density of 1.6 A/cm^2 for 30 to 75 sec, depending on the orientation of the single crystal. For each microsection 30 impressions were made, i.e., the microhardness for each particular orientation constituted the average of 30 readings.

The molybdenum single crystals grown preferentially in different crystallographic directions had different tensile strengths and ductilities:

	[110]	[100]	[112]	[111]
σ, kg/mm^2	40.6	43	44.9	49.9
ε, %	18.5	14.6	20.3	9.4

The maximum tensile strength occurred for single crystals with the [111] orientation, the value being 20% above that of other crystallographic directions.

The differences in the ductility of single crystals with different orientations correlate with the values of the relative elongation and the different character of the rupture mechanisms and stress/strain curves. The greatest ductility occurs for single crystals grown in the [112] and [110] directions. The elongation of such single crystals at rupture is 1.5 to 2.0 times greater

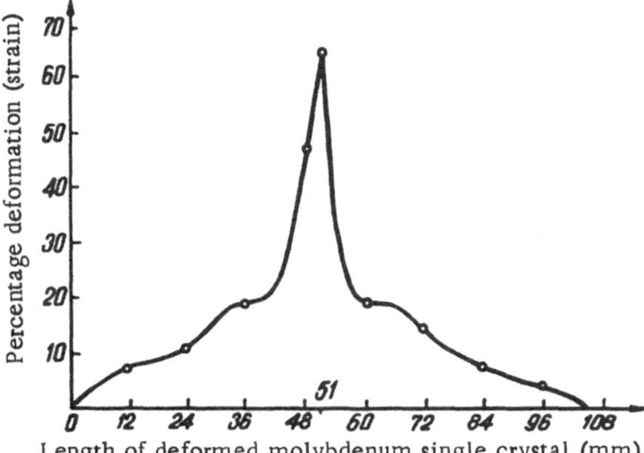

Fig. 1. Linear distribution of the deformation suffered by a single crystal grown preferentially in the [112] direction.

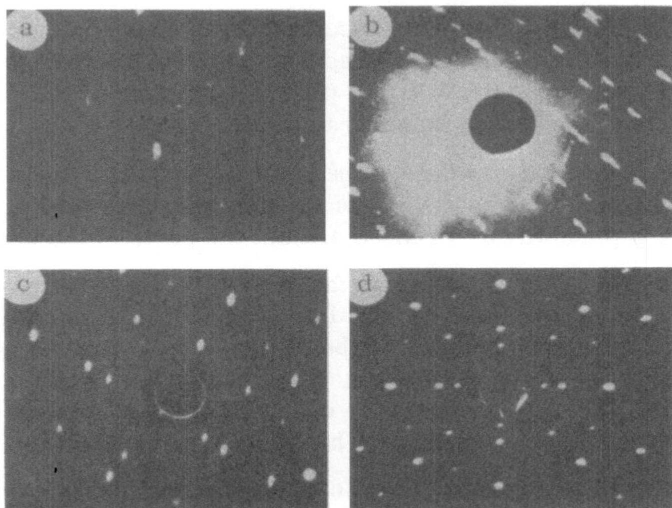

Fig. 2. Laue photographs representing the rupture of single crystals with the [110] orientation: a) at the point of rupture; b) at a distance of 30 mm from the latter. Also analogous pictures for the [112] direction: c) at the point of rupture; d) at a distance of 30 mm from the latter.

than in other directions. The anisotropy of the mechanical properties of single crystals with different orientations is borne out by the microhardness measurements. In single crystals grown in different directions, the shape of the impressions differs from one to the other, and the hardness values also differ appreciably. Single crystals with the [111] orientation have the greatest microhardness (215 kg/cm^2), 10 to 11% more than in other crystallographic directions.

Under tensile strain, single crystals with the [112] and [110] orientations ruptured by a shear mechanism, with a preliminary contraction of 60 and 45% respectively, followed by rupture, while single crystals with the [111] orientation ruptured by way of shear in two mutually perpendicular directions. The distribution of the degree of deformation over the length of the most ductile single crystals oriented in the [112] direction is shown in Fig. 1.

The Laue photographs taken from a section parallel to the rupture show that rupture occurs in a plane close to the (100). For single crystals oriented in the [110] direction there is a characteristic shelf formation, comprising two planes of the $\{100\}$ type situated at an angle of 90°. The cross section forms an ellipse, the major axis intersecting the $\{\bar{1}00\}$ plane and the minor axis the $\{100\}$.

The asterism on the Laue photographs obtained from the rupture point (and points close to this) indicates a substantial change in structure (Fig. 2). At a distance of 30 mm from the rupture point this structural disruption is in no way visible.

Analysis of the asterism by reference to the Laue photographs shows that, in crystals with [110] and [112] axes, slip occurs along two perpendicular systems of planes of the $\{110\}$ type in the $\langle 111 \rangle$ direction. For example, in the case of one of the single crystals with an axial orientation [110], slip occurs along two planes, the $(\bar{1}01)$ and the (101).

References

1. E. M. Savitskii, Ch. V. Kopetskii, A. I. Pekarev, and M. I. Novosadov, Zavod. Lab., No. 8, p. 1041 (1961).
2. E. M. Savitskii and G. S. Burkhanov, Pribory i Tekh. Éksperim., No. 4, p. 248 (1965).
3. E. M. Savitskii and G. S. Burkhanov, Metallography of Refractory Metals and Alloys, Izd. Nauka, Moscow (1967).

PHYSICOMECHANICAL PROPERTIES OF DEFORMED MOLYBDENUM SINGLE CRYSTALS

G. A. Klein, S. M. Mikhailov, L. Kh. Osipova, V. E. Grakov, O. V. Volkov, and M. G. Sultanova

In this paper we shall consider the changes taking place in the physicomechanical properties and structure of molybdenum single crystals as these undergo deformation in the course of forging and wire-drawing. The samples for study consisted of molybdenum single crystals grown in specific crystallographic directions, [100], [110], [111], and obtained from metal powder by the method described earlier [1]. The single crystals were subjected to various degrees of reduction (50, 75, 90%) in the course of forging.

Before deformation of the single crystals, the crystallographic orientation of the latter was verified by the Laue x-ray method. It was found that single crystals of different orientations had deviations of 3° from the crystallographic axis for the [110] and 5° for the [111] directions. Of the resultant forged bars, some were tested for strength and ductility and some were subjected to a wire-drawing process, the degree of reduction at each pass being 5%. The forged rods intended for mechanical testing were first vacuum-annealed at 400°C and $1 \cdot 10^{-5}$ mm Hg and then ruptured on a universal screw-type machine designed for tensile and compressive testing (type UM-5) furnished with an automatic recorder.

Wires of different diameters were also tested for strength and ductility. In order to discover the influence of annealing temperature on the rupture characteristics of the wire obtained from single crystals of different orientations, the wire was annealed in vacuo and in hydrogen at temperatures between 500 and 1350°C.

The change in the structure of the single crystals in the course of forging and wire-drawing was determined, after the mechanical polishing and chemical etching of the microsections in a mixture of potassium ferricyanide, caustic soda, and distilled water (ratio 1:2:18), by microstructural analysis in an MIM-8 metallographic microscope and also x-ray diffraction (back-reflection x-ray photographs).

Figure 1 shows the tensile strength of deformed molybdenum single crystals in relation to their preferred orientation. The deformation (75 to 90% degree of reduction) of single crystals oriented in different ways tends to strengthen the molybdenum. However, the tensile strength increases to different extents in the case of single crystals of different orientations. The greatest increase in tensile strength (almost 60%) occurs for the [110] direction and the smallest (30%) for the [100] direction.

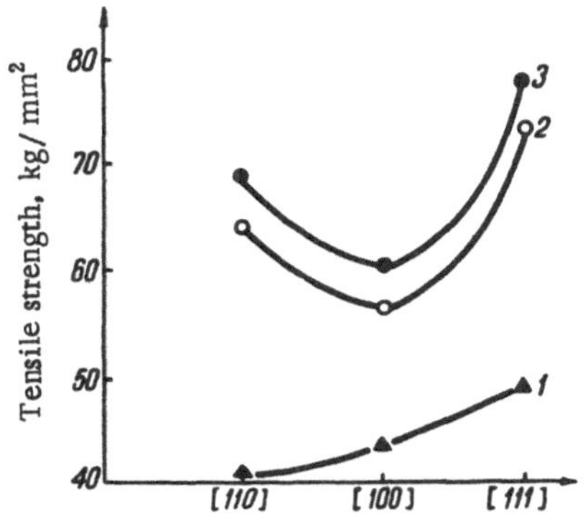

Fig. 1. Tensile strength of molybdenum single crystals as a function of orientation and degree of deformation: 1) undeformed; 2) deformed 75% (by forging); 3) deformed 90% (by forging).

The ductility of the deformed single crystals also depends on the corresponding orientation [2]. We found that the smallest ductility (as indicated by the elongation at the point of rupture) occurred for single crystals oriented in the [111] direction on deforming by 75 and 90%. The microstructure of the deformed single crystals indicates the occurrence of a complex process of deformation in the molybdenum, leading to its hardening.

On considering photographs obtained from microsections of deformed single crystals grown in the [110] direction, we found that a shear motion took place over the slip planes under the action of the deforming stress. After the deformation of the crystal in the course of forging (50% reduction), slip bands could easily be traced. On increasing the degree of deformation to 75%, the number of bands increased, their density became greater, and the distance between them diminished. On subjecting the single crystal to 90% deformation, a fine-grained polycrystalline aggregate developed.

There is also a difference in the microstructure of deformed single crystals of different orientations; in single crystals oriented in the [110] and [111] directions and deformed by 75% there are clear slip bands on the background of the single-crystal structure, running uniformly throughout the whole microsection and situated at short distances from each other. In single crystals with the [100] orientation, slip bands are less visible.

On subjecting the single crystals to 90% deformation, fine-grained polycrystalline aggregates are formed for all the orientations under consideration. On annealing the single crystals deformed by 75 and 90% at 400°C, the recrystallization process takes place in a manner depending on the original orientation.

X-ray structural analysis of the deformation of single crystals showed that, after 50% deformation, the x-ray diffraction pattern exhibited uniformly-darkened Debye rings, consisting of well-resolved point interferences; this corresponded to the break-up of the single crystal into grains forming an asymmetrical structure during the forging process. On increasing the degree of deformation to 75% there was a more uniform darkening of the Debye rings, indicating a further reduction in grain size.

The x-ray diffraction patterns of single crystals deformed by 90% show uniformly-darkened Debye rings without any point interferences. Thus a further break-up of the single crystal into very fine grains occurs in the third stage of forging.

In tensile rupture tests, deformed single crystals with different orientations exhibited a typically ductile fracture with the formation of a neck in the middle of the bar. Photographs of the rupture points taken along deformed single crystals of different orientations showed that, in the tensile testing of 75%-deformed single crystals, the slip lines were mainly situated along the surface of the bar, whereas, in the case of 90%-deformed single crystals of all the orientations studied, there was a uniform array of slip lines, extending up to the point of rupture.

Thus the results obtained by light microscopy and x-ray structural analysis agree excellently with the results of mechanical tests, indicating a direct relation between the degree of deformation and the strengthening of the oriented molybdenum single crystals.

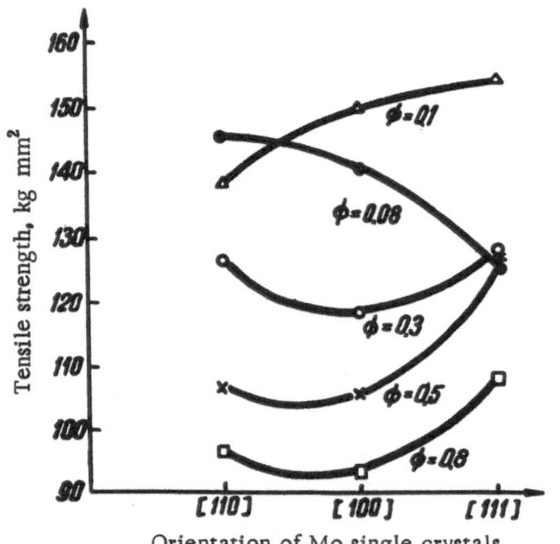

Fig. 2. Tensile strength of wires of different diameters drawn from Mo single crystals of various orientations.

In the later stages of deformation (cold wire drawing), a further increase in strength occurs. The tensile strength and ductility of the wires depend not only on the diameter (degree of deformation in drawing) but also on the crystallographic direction in which the original single crystal was grown (Fig. 2). This orientation dependence (anisotropy) is preserved in wires up to a diameter of 0.8 mm, and also after annealing, right up to the recrystallization point. The maximum ductility (elongation) of wires prepared from single crystals grown in the [111] direction was obtained after annealing at 1050°C, and that of wires prepared from single crystals grown in the [100] and [110] directions after annealing at 1000 and 950°C respectively.

The microstructure of the microsections of wires annealed at the temperatures in question shows that the sinuous structure characteristic of unannealed wire vanishes and primary-recrystallization grains are formed.

References

1. G. A. Klein, M. S. Mikhailov, L. Kh. Osipova, V. E. Grakov, and G. N. Grishkov, Production and properties of oriented single crystals from molybdenum powder, this volume, p. 138.
2. N. V. Ageev, A. A. Babareko, G. E. Chuprikov, and N. N. Babareko, Izv. Akad. Nauk SSSR, Metally, No. 4, p. 84 (1966).